THE DARK SIDE 2

DAY ZERO AND OTHER STORIES

by Adrian Tilley

QX PUBLISHING CO.

The Dark Side 2: Day Zero and Other Stories

Author: Adrian Tilley

Editor: Betty Wong

Cover Designer: Tina Tu

Published by:
QX PUBLISHING CO.
8/F, Eastern Central Plaza, 3 Yiu Hing Road, Shau Kei Wan, Hong Kong
http://www.commercialpress.com.hk

Distributed by:
SUP Publishing Logistics (H.K.) Limited
3/F, C & C Building, 36 Ting Lai Road, Tai Po, N. T., Hong Kong

Printed by:
Elegance Printing and Book Binding Co., Ltd.
Block A, 4/F, Hoi Bun Industrial Building, 6 Wing Yip Street,
Kwun Tong, Kowloon, Hong Kong

Edition:
First edition, April 2019
© 2019 QX PUBLISHING CO.

ISBN 978 962 255 134 3

Printed in Hong Kong.

Introduction

To Young Readers

Welcome to my third volume of short stories. The first, *Cheung Chau Paradise*, was stories about teenagers' everyday lives in Hong Kong. The second, *The Dark Side* was various horror stories, all set in the city with teenage central characters. This collection – *The Dark Side 2 : Day Zero and Other Stories* – is a mix of stories, dividing broadly between six 'horror' stories and six 'thriller' stories. Horror stories do what they state – they horrify us, fill us with dread, often of our worst fears (pain, death, loss and so on); they show us a much darker side of human behaviour. At the same time, they can be funny and satirical. Having been a teacher, I enjoy creating teacher characters who are ridiculous or sad or even mad. The rituals of school (assemblies, uniform, language) are often absurd and well worth laughing at by plunging them into bloody chaos. It is worth remembering that real-life horror is exposed to us every day in the media: accidents, wars, catastrophes. Horror stories may just prepare us to admit those real stories into our experience.

The other six stories are more varied. Some are thrillers where the main aim is to excite the reader, drive them on into the story wondering what will happen next. Some like 'Auto Message' try to say something about how technology is affecting our daily lives. 'The Pickers' is very different in that it's based on true events when, in 2004, 24 Chinese workers were drowned on the coast of England while working illegally. The story is both horrifying and sad and reminds us that the international slave trade still thrives to everyone's shame.

Hong Kong is a great city to set stories in – it has an energy, a sense of order and a sense of chaos, that lends itself to stories. Look at how many films have used its unique settings to good effect. I hope my admiration for the place shines through and that perhaps the stories will open your eyes to what the city state is: diverse, exciting, full of promise and possibilities.

To Teachers

This collection of stories should appeal to young Asian readers as the characters and settings are (except in one case) all centred on Hong Kong. The English language content is challenging without being too obscure and is more accessible than much of the prescribed literature study on the exam lists. There are also serious issues addressed among the stories: modern slavery, the impact of technology on our lives, fears for the future of our security. Two of the stories are in play form and so should be useful for group/paired reading in class. I hope you will find the stories useful and they will lend themselves to interested reading and engaged responses to the texts.

To Parents

You can't underestimate how important reading stories is in the development of language acquisition and intellectual development for your children. It has a dimension – a strength – that straight language learning cannot have. Learning words and parts of speech and sentence construction is only a small part of language. Engaging the imagination and finding language to express the ideas there, works on a different level of learning. There is also the bonus of working with idiomatic language – the everyday language – that offers a more 'natural' tone to English expression. The other great thing about stories is that they carry the reader into another world (however frightening or challenging or amusing that might be). The reader can experience things – extremes – in a safe way. When things get too scary, the book can always be closed. Reading stories is so rewarding. As Einstein said: 'If you want your children to be intelligent, read them stories. If you want them to be very intelligent, read them more stories.'

Who can dispute Einstein on this?

A few tips on parental reading guidance (I write as a parent and a grandparent):

Read what your children read. Get to know what they like and why they might like it (even if you don't!)

Read with them and to them. Even teens like having stories read to them.

Talk to them about what they're reading. Get them to predict what they think the ending will be. How would the story be different if…

Encourage them to keep a record of all their reading – a journal or diary.

Encourage them to write in response to the stories.

Find other ways for them to respond to the stories. Drawings. Diagrams. Speech bubbles. Thought bubbles. Make a video. Anything is possible.

Be good role models. Read yourselves so your children see it as an important activity.

Good luck! Happy reading!

Adrian Tilley

Contents

Day Zero

Day Zero
I'd read about the 'end of the world'. It had been promised on several occasions. 1669. 1849. 1914. 1987. The man who sold papers and magazines with smiley naked girls, told everyone it was going to be a Thursday in 20 – but that had passed and gone. I think he claimed to have been informed by aliens that this would be the date so I largely ignored his idea. But other, more qualified experts, have suggested that the calamity was close. So it wasn't a real shock or surprise when the newspaper headlines were 'End of the World' (a bit lacking in imagination). And the television news, too, followed the story. The news reader had a big frown and hadn't put on lipstick. So it was serious. So this was it. The End of the World. And I hadn't even taken my exams. What a waste of all that worry and homework and being told off for not taking it seriously. The exam, I mean, not The End of the World. Which is very serious.

No, I take that seriously even if I don't understand. I've actually decided it's so complicated that I can't begin to explain. It just is. A sort of big, black cloud hanging over us (but not actually a cloud. That's a sort of metaphor, my English teacher told me). It. TEOTW.

We'd been given the date. Twenty-Third of April and then

that would be it. I asked my mother and father what 'it' was. They looked at each other, then at me and began crying. I asked my grand-parents – they both lived in the flat with us – and they did the same. I was going to ask my History teacher at school because she knew everything – but school was all locked up. Not a sign of anyone. Not even a notice to say: 'School closed for the End of the World'.

Mother and Father talked about taking all their basic possessions and getting on a train to China and hide out somewhere in the countryside, away from 'It' – you know. But all the trains were cancelled by the government to stop anyone trying to escape. Anyone meant 'everyone'. All the roads out of the country had been blocked and all aircraft grounded.

'We'll just have to stay here and sit it out,' declared Father.

But I don't think you can sit out TEOTW. It's not like having a cold and watching a boring film. It is The End of the World.

Day Zero Plus 1

Things quickly started to go wrong. Lots of people, who, like my father, thought they could sit it out, decided to stock up on tins of food. Soon the supermarkets were being overturned, people just going in and stealing what they wanted. Then it was the K-Marts and the 7-Elevens. People were fighting each other for a bag of bar-b-q nachos. The shop owners were fighting the people fighting – hitting them with anything they could get hold of including bags of bar-b-q nachos. It was mayhem.

I watched from the flat as a police van roved into the street where crowds had gathered to raid one store. The van drove straight past, pulled up outside another store and all the police scrambled out, broke down the door and took out everything they could lay their hands on. They piled the stuff in the back of the van and drove off. Bags of what looked to me like nachos dropped out of the back of the van as they went.

I don't know what it is with Hong Kong people and nachos but whatever it is, it isn't healthy. Too many nachos are very bad for you and cause long-term damage. I suppose with TEOTW with us, it doesn't matter anymore.

Day Zero Plus 2

Electricity was switched off last night – all night – so from 6.30 on, there was no light except for a few candles which we'd found under the sink. I've never seen the city blacked out before – it's always been so bright at night, all the buildings lit up, all the blocks of flats lit up. But this was weird. There was low cloud so no light from the heavens at all. It made everything quiet too – almost silent – except for the breaking of glass where looters were raiding shops and houses. Oh and the occasional gun shot.

Day Zero Plus 3

All television and radio services are closed down. There

is no internet, no mobile networks. Nothing. We no longer know what is happening except what we can see from the windows of the flat. Father says at breakfast (yes, we still have enough stuff for a usual breakfast – ham and noodles) he is going out to find out what is happening, maybe join up with some neighbours – 'strong together' he says. Mother laughs at him, calls it 'nonsense' then breaks down into tears again. Father takes a knife from the kitchen drawer and slides it carefully into his belt. He looks like one of the pirates in my old picture book (but no eye-patch or parrot).

As he goes, I slide out and find my way to the roof of our block of flats. It has a notice that says 'No Entry' but the notice has been pulled off and the door's lock mechanism has been broken. From the roof you get a good view out across Yau Ma Tei to Jordan and the Harbour. There is very little noise – there is hardly any traffic – and there are lots of plumes of smoke rising from different places all over the city. Perhaps it's accidental fires or maybe naughty people have set light to things. I don't know.

I watch Father from above. He's standing outside the entrance with one of his friends from the floor below. His friend is carrying a walking stick upside down. And he's not using it to help him walk. They cross the road looking both ways as if a 15B bus will suddenly appear and run them down. There are one or two men on the street, sort of hiding in shop doorways. A mother, pushing a pram, is running from the top of the street past our flats. When she gets level, I see her pram doesn't have a child in it, but a television set. Seems strange to be taking a TV set for a walk (or in this case, a run). At the other end of the street, two young boys are throwing large stones at a parked car. They do it without speaking as if it's a job they have to do. The car windscreen caves in first, then the side windows shatter. With all the windows broken, the boys move on to the next vehicle and begin again, silently throwing rocks. No-one comes out to stop them.

Father and his friend have crossed the road and are peering into a locked-up shop which is usually 'Hearty Cake Bakery'. I wonder if they're looking at a tray of custard tarts but then remember the shop is not making or selling anything. No Hearty Cake for anyone.

That's when a red mini-bus suddenly roared round the corner, nearly tipping over. Its front number plate is dragging on the ground making a screeching sound and sending up a shower of sparks. It shudders to a halt outside the bakery, the side door slides open and a gang of young men jump out, shouting and waving knives and sticks. Before Father and his friend can get away, the gang are on them, beating at them with sticks. The two men fall and try to crawl away.

I shout, 'Father! Father!' then, 'Stop it! Stop hurting him!'

But maybe they don't hear me. Or the gang don't care.

I begin running back from the roof and down the stairs, taking as many steps as I dare at each jump. It is ten flights of stairs and my legs are burning by the time I get to the bottom. I burst out onto the street and run towards the bakery. My father is still crawling away but his friend is behind him being kicked by two of the gang. As I draw close I can see the friend's head doesn't look right anymore – it's at a funny angle.

My father looks up as I get to him and puts his hand up. Does he want me to stop or help him? I put my arms under his and drag him to his feet. I look behind him and the gang are now only interested in breaking into the bakery. In seconds they are running out with bags of flour, sugar, dried milk and throwing it into the mini-bus. They then scramble back into the bus and it lurches off. I pull Father to one side as it roars past, inches away.

Is this what the End of the World means?

Day Zero Plus 4

Father has been lying in bed since yesterday, recovering. He has a lot of bruises and tells us he thinks he has a broken rib. His face is swollen and there are cuts and grazes on his arms and legs. Mother looks after him carefully and tells him what he can and cannot do. He cannot go to the toilet so I am now 'toilet monitor' and bring him an empty bottle from time to time to fill up. I also bring him a bottle of water to drink. I tell him maybe it would be easier if, instead of me taking away his filled bottle, I leave it for him to drink. A simple re-cycling process. But he doesn't like the idea, laughs at me then swears because it hurts his ribs.

Mother is really upset with all this and now locks all the

doors in the flat, all the time. 'Just in case.' The food is running low already and water is now only dribbling from the taps so we are collecting as much as we can. None is going on washing anything – clothes, dishes, us. No, nothing. We must have it only for drinking and some cooking. Mother doesn't know what to do about getting more food. She is terrified of gangs that may be outside. It leaves only one thing: I have to go out and steal some food. It's just that I won't tell her. Father can't do anything anyway. Tonight I will go out.

Mother and Father are both asleep. Father's noise is like bubbling breath. He is not looking good. I pull on my soft shoes, black jeans and a black t-shirt, pull a back-pack on, then pad over to the front door. I take the key and unlock it, slip through silently and pull it closed. I use the key to turn the lock back. I stop and lean against the door. Maybe this is too much and I should go back. Stupid girl, thinking you can go out and steal. Steal what? From where? There is a bumping sound along the corridor. It is so black I can see nothing but I know something is coming my way and quickly. I dive left and move to the stairs, through the door and hide down behind the stairs. The bumping noise stops at the stairs. A flashlight goes on and I see our neighbours – Mr and Mrs Wong – their eyes are wide with fright and I see why. They are carrying a body between them and the head is banging against the wall as they shuffle along. Then I see in the pool of light that the body is Mrs Wong's father, who lives with them. Lived with them. He is obviously dead. Are they taking him out to bury him?

The lamp goes out and they drag the body through the doorway but I am gone. Down the stairs like a ghost. All the way down. Throw myself at the door latch and I'm out into the street.

There is a strange smell in the air: sweet, sickly, it catches in my throat. Then it's stronger – the stink of death? Rotting flesh, petrol, sour milk – it's as vile as it gets and I want to throw up. But no time. I pull a handkerchief out of my pocket and tie it round my face.

I head off right, past the cars and vans parked at funny angles in the road. When did people learn to park like that? The street looks deserted but in the darkest corners and alley openings, I feel dark shadows watching me.

'Hey, boy! Come here!'

Three heaps of darkness shuffle together from the building opposite. I decide not to explain that I am actually a female and now is not the time for chat-up sessions. I sprint right and hear them shouting nasty words behind me. Really nasty words. And they're not shuffling anymore. They are running.

Now, I am a bit skinny. Everyone said so at school. And I can run because I won the Athletics Cup last year at the sports stadium in Sha Tin. It's funny to think that will never happen again. No more cheering and house colours and announcements that no-one can understand. Never again.

These thoughts don't slow me down luckily and I hare off down streets I know so well, even if it is pitch black.

But you can only be clever like that for so long. I turn into Reclamation Street and there at the end is a gang of men sitting around a fire piled up in the middle of the road. They haven't seen me yet but they will. I dive behind a big, black van. Sure enough the three heaps arrive at the turning but they're not heaps any more. In the light from the fire which bounces all the way down the street, I can see they are three young men with clubs and meat cleavers in their hands. I don't think they're going to a bar-be-cue either. Now they are walking towards the fire, perhaps trying to figure out who is there. Perhaps they've forgotten about me. As they near, I slide along the van. I reach for the door handle and the door clicks open. I crawl in, ducking low across the front seat.

The three men are still advancing on the fire gang who number about six. I now see they are all women – well, girls – all with scarves wrapped round their heads. They've used lip-stick to paint lines and circles on their faces. In the light of the flickering flames they look…well, mad. They are crazies. They've seen the three males and stretch out in a line in front of the fire, now black outlines who have stepped from the flames of Hell.

'They're all girls,' laughs one of the men.

'Clear off, girls. This is men's country,' laughs another.

One of the girls steps forward.

'What did you say?'

Her voice is sharp and prickly like cactus.

'I said, 'This is men's country,' repeated the male.

The girl's arm moves so quickly I almost don't see it. The young man shudders, then drops to his knees, looking down at the knife sticking out of his stomach. He screams. His two companions turn to help him and the girls are on them, stabbing, hacking, whirling, shouting.

I slide down into the driving seat. I can't watch. All I hear is the screams, the pleadings, the thuds and the girls' laughter. The next time I peer over the windscreen, the men are three dark shapes again, sprawled on the tarmac on top of a dark and growing slick of shiny stuff. The girls are back at the fire, laughing and slapping each other. I have to get away from this hell, I think. I don't want to see any more of this. My father's friend. Now these three – even if they were after me. I wasn't expecting that.

There is a door in the back of the van cab. I open it and crawl in. The light from the fire spills in too and I'm surrounded by white bags. Bags of tapioca. Perhaps they've been forgotten or left here. Perhaps the van belongs to the gang of girls. I stash as many packets of tapioca as I can into the back-pack, lash it tight and feel my way to the back of the van. I open the door and slide down to the road. Bent double I move along the row of cars and vans, all abandoned it seems, not daring to look back at the girl gang. They would be on me in seconds and I would be just another sticky, dark shape in the street.

I reach the corner and dive right. No shouts. I've not been seen. The girls are too busy re-living their violent glory.

'Tapioca? Tapioca? Where did this come from?'

Next morning my mother is weighing the six white bags in her hands.

'Who got this?'

I look at my feet.

'The Tapioca Fairy.'

My mother dumps the bags on the kitchen table.

'You? You got these? You stole these?'

Now I look at her, pleading.

'I found them in an abandoned van. By chance. Good luck, eh?'

'When?' Her eyes are large and staring.

'Last night I went out. Foraging.'

Foraging was a new word I'd learned at school when we did primitive Chinese tribes in History. It was a much better word than 'stealing'. I was watching Mother intently, waiting for the volcano to explode, waiting for her hand to smack my face. Then her arms are around me. She is sobbing again. Her tears are in my hair.

'I don't want to know, my child. I don't want to know anything. But you have saved us.'

I look up at her.

'And we won't die of eating too much tapioca?'

Five more days have passed …

Day Zero Plus 10

It is hot and sticky all the time. No electricity means no air-conditioning. If we open the windows, the death smell comes in. The fires have spread across the city and a layer of smoke lies over everything. Ash and dust hover in the air all the time. I spend a lot of time watching from the roof. I see other people on other roofs but no-one waves. Everyone just stares.

No-one goes onto the streets. They belong to the gangs. I saw the girl gang go by and I automatically ducked down in case they saw me. Even though I know they hadn't seen me. Sometimes an armoured truck rolls down the street. The gangs throw stones at it but it's all pretty pointless. I think everyone knows that now. They're all moving slowly as if to save energy.

Father isn't recovering. He's grey now – his face, his hands. He seems to be shrinking, day by day. I go and sit with him. He carefully wipes the dust from my hair, my face and smiles. Talking is a big effort for him.

'Ah, my daughter. Such a shame.'

His hand slides off me and he falls asleep again. Is it a shame I'm his daughter? No, I suppose he means this. All this. The End of the World stuff. I'm glad he sleeps because he doesn't see me cry. And he can't hear Mother cry. Which she does most of the time. Perhaps it's the tapioca diet. Probably it's just that everything has got on top of her and is driving her, like a rusty nail, into the ground. She spends a lot of time just sitting and staring. Which is another reason to go up on the roof.

Then even that isn't possible. Today, I am sitting on the ledge, looking down ten storeys to the children's playground below. I hear a thud in the brickwork by my hand. The wall shudders and chips of brick spray up. The bullet aimed at me wails away into the sky.

Before the second bullet hits the brickwork, I have fallen off backwards (not before I give a sickening jerk forward which nearly sends me plummeting to the concrete below.) My stomach and knees turn to water. Someone's trying to shoot me. Even squatting beneath the parapet I can see the roof of the next block of flats. An old man is sitting reading a book. So calm, when the world around him is falling apart. I want to shout at him, to warn him someone has got a gun but, before I can. The book he is holding explodes into a dozen white butterflies. He sits bolt upright, his back against the wall. There is a troubled look on his face as if he is trying to work out some big mathematical problem then he slowly slides sideways. The wall behind him is smeared dark red.

Who would do such a thing? Who could just do that? Shoot people because they are there. Not stealing. Or fighting. Just quietly being there, like porcelain targets. This is what the End of the World means. The end of thinking. The end of Reason. Everything is madness. Illogical. The end of imagining what Life is, what Life means. A nothing is taking over. A savage, violent nothingness. Because nothing matters anymore.

Ten more days have passed …

Day Zero Plus 20

Looked in the mirror today. I hadn't meant to. I avoid it. Catch sight of this dreadful creature with a filthy face, greasy hair hanging down like rats' tails and a skinny body with ribs sticking out like a rain grating. Who is this thing? I feel sick with the shock. This is me. This is how I am now? A stick person. A forked thing like a snake's tongue. I don't look like a human any more.

Everyone is hungry – all the time. Not just hungry – I mean – obsessed with being hungry. All we think about, all we talk about is food and where we are going to get it. We made the tapioca last till yesterday. Now it's just noodles which we fry. The water nearly ran out too but then it rained. Everyone rushed onto the roof or into the street with pots and jars to collect the rainfall. There were arguments and fights and some people had their water stolen. But at least we have enough drinking water for the next few days. We sip water every hour. It's the only way. Except we allow Father more – he's desperate for water all the time. He could drink Mother's tears.

But it's the hunger that drives you mad. All the time – food, food, food. Talking about meals we've had. Meals we'd like to make. On and on. And the gnawing pain in the stomach burning you up.

I look again in the mirror. Closer this time. My face is covered in small spots – all our food is now cooked in oil. I have some boils on the back of my neck which are very painful. The skin on my hands is hard and flaky. My tongue is yellow and coated in…it feels like a lump of fat foam in my mouth. My throat aches all the time. I think my teeth are coming loose.

I'm falling to bits here. Slowly falling to bits. It's like I'm a jigsaw and someone is shaking me, breaking me.

Nothing seems funny or odd anymore. I've stopped laughing at things. Everything is changed. Everything. This me is a different 'me' – it's not a 'me' that I like. At all. It's a 'me' that is ugly with sharp points and sharp words. I shouted at my mother today for crying all the time. What is the point of crying?

One more day has passed…

Day Zero Plus 22

Father died today. I howled and howled and howled. My mother shrank even smaller. Her tears leak from her like a tap. She is leaking her life away.

We know we can't keep his body here. Some neighbours come in and we light candles, then wrap his body in plastic and drag it out into the corridor and down the flight of stairs (the lift hasn't worked for a long time). The body is dropped several times – it is awkward on the corners – but it is light. There was nothing of Father when he died – just a bag of bones lying in the bed.

As with all bodies, it is the family responsibility to burn the body. All the graveyards are full and disease-ridden (some people just dump bodies there). The neighbours help gather up some rubbish and broken furniture to make a pyre. Someone has some petrol siphoned from an abandoned car and it is liberally sprinkled on the wrapped body and the pyre beneath. Someone else offers Mother a match to light the fire but her hands tremble so much she drops the box. I pick it up, slide out a match, strike it and drop the flaming little stick onto the stinking pyre. There is a dull thud and flames leap up. Mother collapses against me and I hold her up. Somewhere from within my brain, some prayers appear and I shout them at the roaring flames. Smoke turns black. Rubbish and wood crackles. Tears fall – enough to drown the world.

Day Zero Plus 23

My headache returns again but there are no pills. No relief. I know I must go and get food. All the shops locally have been emptied by the looters but I remember that there is an area of sheltered ground in the old West Kowloon Park that some people use for growing their own vegetables. It might still have some edible stuff there and not many know about it.

It's a long walk and my energy levels are low. I have a rucksack on my back and a kitchen knife in my belt (and not just for cutting vegetables). The pathway leads over the old motorway near the Western Harbour Tunnel. Now the motorway is empty of traffic. No more flashing lights and warning signals. A strange silence. There is smoke coming from the huge shopping mall to my right. I can see all the glass entrance is smashed. But there is no-one around.

The West Kowloon Park is wild and overgrown. There had been plans for theatres, cinemas, restaurants – but none of it had happened and the area had been abandoned. A few locals had occupied the area and cultivated crops and that was what I was looking for now. Hiding behind a dumped, rusting mini-bus, I can see rows of onions, carrots, even some potato plants in neat straight rows. They are being looked after still. A bag full of that stuff will last us weeks. I survey the area carefully. At the far end of the growing area, two women are hoeing and weeding, their backs to me.

I slide forward on all fours, keeping low to the ground and scramble towards the first rows of onions. The soil is dry and loose as my fingers close around the hard, round vegetables and stuff them into the bag which I have by my side. I take half a dozen then move towards the bright green, frilly tops of the carrots.

'What the hell do you think you are doing?'

The hard voice comes from above me. An old man – he is tall and thin in a t-shirt and dirty trousers – is standing there, pointing at my bag. His other hand grasps a spade. His little, furrowed eyes burn with anger.

'I'm not…I'm…I'm just…,' I mumble.

I'm raising myself from my knees and squatting beneath him. My hand is reaching back to get the knife. My fingers wrap around the handle.

'You thief! You're stealing my food! That's mine!'

He half-turned to call to the two women a hundred metres away.

'We've got a thief! A bloody kid thief!'

He turns back to me and raises the spade above his head. It will split my head like a meat cleaver. I know that. From my crouching position, I spring upwards at him, thrusting the kitchen knife in front of me. It thuds into his chest, the blade breaks off and just holding the handle, I fall across him somehow, knocking him off-balance.

He lets out a painful yell and looks down with a surprised look on his face at the shiny metal sticking between his ribs. He drops the shovel and falls back into a sitting position. His hands scrabble at the knife blade, trying to get a grip but it is already wet and slippery with blood.

He curses me and curses me, over and over but I am deaf to it all. I can see the two women hurrying closer, shouting. This is not the way the mission should end but I must complete. With the old man groaning in pain, then shouting at me, I grab handfuls of carrot then switch to the potato rows, piled neatly, so when I tug at the green tops, the white nuggets of food roll out onto the black soil.

The women are fifty metres away. The old man is reaching for the shovel again but his hand seems to have trouble gripping it. I have the rucksack full and he is screaming at me. I try to say something.

'I'm sorry…my mother…'

Then I run and soon his shouts and screams disappear as I'm back over the motorway, heading home.

Night. Can't sleep. Did the old man die? I didn't tell Mother where the food came from. She probably thinks I'm out prostituting for it and didn't want to ask. Apparently that's what lots of the girls do to get food. More likely, Mother doesn't even think where the food comes from.

Now I'm a thief. And perhaps a murderer. I am no better than all those other animals on the street. Maybe – but perhaps I don't care anymore. The fried onions and potatoes fill my stomach and we're going to live for a few weeks more. My headache has gone. Mother has stopped crying. Just for a time.

Seven more days have passed…

Day Zero Plus 30

Am looking through old school books. Old? How long ago was that when schools were open? All that time spent on learning, remembering, testing. All a waste of time because there's nothing to learn now. Except how to steal food and how to defend yourself. We keep the front door locked and barricaded with furniture as we've heard gangs are looting apartments now. I keep a knife in my belt all the time.

I'm trying to remember how it all started. The End of the World. Some said it was natural virus that spread like wildfire across the world. Others say it was poisons rocketed into the air by North Korea. Whatever it was, it made everything collapse. First the hospitals which couldn't keep up. Doctors and nurses died. Medicines ran out. Then all the services – electrical power, transport. Then the banks. And that was it. No government except some armed police and soldiers on the streets but then they formed separate militias and occupied different parts of the city – their own territories with flags and everything. Just chaos really. So easy how a complicated people world comes crashing down and is reduced to smoking cinders and street fighting.

I put the school books in a neat pile and carefully lift them. I walk slowly to the window and launch them into the air. They flutter like butterflies.

Food is low again. I'm going to need to go out again. Is it worth it to try the vegetable field again or should I start making rat traps like everyone else? Apparently the meat of the rat is rich tasting. I think that can wait. I'll see what there is in the garden. I hope the old boy didn't die but I don't want him to be there either. He'll probably be pretty mad at me.

Zero Day Plus 31

Yesterday the vegetable plot was empty. Everything had gone it seems. I went into one shed – it was humming like an electric fan was on but when I opened the door it was millions of flies feeding on a corpse. I didn't try to see if it was the old man. The stink was disgusting. I slammed the door shut. In another shed, I got lucky. Hidden under some old sacks were bags of root vegetables. They were going off – mouldy and soft – but if they were boiled up, they'd make a great soup, enough to last for a few days at least. Scrambling home over debris and rubbish in the streets, dodging and hiding from anyone who might steal my 'find', my spirits were almost lifted.

It wasn't to last for very long. When I got back to the apartment, I could see something had happened. The front door was hanging, splintered on its hinges. I rushed in shouting for my mother. The large kitchen window was broken. There was no sound in the flat. No sign of Mother. I dare not go to the window – its frame was hanging loose in the gap. I daren't look down below – terrified of what I might see. I swallowed hard and my head was numb.

'Mother? Mother?'

I moved slowly, step by step, towards the shattered window.

'Mother?'

'Okay. Time's up.'

The English teacher's voice is bright and encouraging.

'That's all the time you've got for your 'End of the World' story. Well done. Some of you wrote a lot.'

My hand is aching. I put down my pen and lean back in the seat, briefly closing my eyes. I've been in a dark, dark place for the last couple of hours and it's good to be back in the light again. I look out of the window. The sky is clear blue from horizon to horizon, spoilt only by a vapour trail unzipping the sky.

'Wow! Look at the speed of that jet!' someone says from near the window. The teacher inclines her head to look. All the other faces turn, like flowers to the sun.

'That's too fast for a jet, I think.'

There is a note of unease in the teacher's voice.

All eyes now fix on this white arrow in the blue. Then everything changes. There is a flash. A bright flash, a flash that hurts your eyes to make you blink. And the vapour trail dissolves into a huge black cloud that grows and grows, out of control.

'What the hell is it?' someone yells.

I look down at the story pages on my desk and my chest tightens, my heart beats quicker. I feel as if the top of my head is lifting off. I am going to faint. I open my mouth to speak but no words come out. No words at all. No words left.

First Day at School

Mr. Li smiled to himself. He was very happy. This was to be his first day in a new job – teaching at the St. Olaf's School in Kowloon. Stoker School – an international school that took in students from all over the world. At the interview, the Head Teacher, H.E.L. Singh (he remembered the name on the man's office door) had said they wanted 'new blood'. Mr. Li was that new blood. He'd show them how good he was. How he could dedicate himself to the school (a job made easier by being given a flat to live in on school premises). He'd show them his knowledge of Shakespeare, quadratic equations, quantum physics. There was nothing Mr. Li didn't know something about. And he would pass that knowledge into the pupils little, unformed, empty brains. Easy peasy Jackaneezee (he'd read that in an English magazine somewhere).

Although it was a bright, sunny Hong Kong day outside, as soon as he stepped across the threshold, he noticed the dark…and the cold. Why were all the blinds down, the air-conditioning on so high? Through the gloom he could make out a figure in the hallway – Mister Singh.

'Welcome, Mr. Li. Your classroom is this way. Follow me.'

Singh took huge steps and Li found it hard to keep up, clinging as he was, to his leather bag full of registers, plans, mark books, pens, board markers…Up two flights of stairs, another corridor, door after door passing in a flash. How would he ever find his way back?

'Here we are. Class 5 AB negative.'

Singh's hand pressed against his back and he felt himself propelled into the classroom. The door clicked closed behind him.

'Enjoy!' shouted Singh through the keyhole apparently but was that directed at him, Li, or the kids he could see in front of him? Their faces pale circles, watching his every move through the murk.

'Good morning, Class 5. My name is Mister Li.'

He had plastered on his 'serious, don't mess with me' face he'd learned watching an old 1950's American film about a rough school.

Silence. Twenty pairs of eyes were fixed on him.

'Mister Li – a mystery,' a voice came from the back. A few sniggers then silence.

'Very good, whoever that was. A rhyme and a pun. Well done!'

'Like steak.' The voice again. Mr. Li still couldn't locate it.

'Sorry? What do you…?'

'Meat. Well done.'

Mr. Li forced a laugh but it was dry and crackly.

'Another pun! You are very good with linguistic play.'

That phrase would impress them.

'It merely serves a phatic linguistic function,' came the voice again.

Mr Li's voice rose nervously.

'It does, does it?'

'You are aware of Crystal's work, I assume.'

Clouds of confusion were forming in Mr Li's brain. This isn't how it was meant to be. He would try to get back on track with a pun of his own.

'Ah, but the stakes are high.'

The voice rumbled on, now joined by several others.

'But we don't like stakes. We don't like stakes.'

Mr. Li raised his hands to quiet them and forced a smile.

'You all like puns, obviously. But first, a register.'

He took out the book and, with a flourish, waved a pen from his pocket. He looked down but the names were obscure in the semi-darkness.

'Ha! I need some light.'

He strode to the window and tugged on the heavy metal blind. There was a volley of screams.

'Nooo! No light!'

Two boys crashed against him, grabbed the release cord and lowered the blind again. The searing white light disappeared.

'There can be no light, Mr. Li. It is about saving energy. School policy. Bring Mr. Li a candle so he can see.'

The boy who had crashed against him seemed to be in charge. He was tall and handsome, unlike any fifteen-year-old boy he had ever met before. Another boy, smaller with glasses, was already at the teacher's desk, lighting a candle.

'Thank you,' said Mr. Li looking directly into the bright blue eyes of the 'leader'. And you are…?'

'Jonathan.'

'Thank you, Jonathan.' Mr. Li spoke the three syllables of the name separately as if learning it by heart from another language.

Mr. Li scanned the names on the list. A truly international class. Several 'Vlads' (obviously of Slavic background). A Harker or two (Jonathan was one). Bathory (two girls). Le Fanu (must be French). Nos Feratu (Italian or Spanish he wondered).

'Well, what wonderful names. You must come from all over the world.'

Jonathan stood up.

'Mostly Romania. The Carpathian Mountains.'

Mr. Li shook his head. 'Well, I never. How interesting. I must find out more about your country. Get a taste of what it's like.'

Jonathan smiled weirdly. 'It may get a taste of you, Mr. Li.'

A loud laugh went up. Mr. Li joined in, not quite sure why. The boy stepped forward.

'I'll do the register, sir. I know who's here.'

And before he could protest, Jonathan had stabbed a list of red ticks down the line of names. He returned to his seat, his face upturned to the teacher.

'Well, thank you for that…Jonathan. Now what is our favourite subject? You must have one.'

'History, Mr. Li.' Jonathan again – clearly the spokesperson since no-one else challenged with an answer.

'Ah, good!' responded Mr. Li, rubbing his hands. 'Who built the Palace of Versailles?'

'Louis the Fourteenth!' A chorus of voices.

'In what year was the battle of the Somme?'

'1916.' All the voices chimed.

'Right. Good.' This wasn't going quite as he wanted it, so Mr. Li dug deep into his history chest. 'Which three men wrote the American Constitution?'

'Benjamin Franklin.'

'George Washington'

'James Madison.'

Mr. Li gulped and plucked out another question.

'And who was the oldest?'

'Benjamin Franklin.'

And so it went on. The deeper Mr. Li plunged into his personal archive of obscure facts, the quicker the students responded. In the end he had to concede defeat.

'Well, you certainly know your history! It's almost as if you were there at all the events.'

A trail of titters went round the room. Jonathan stood up.

'Yes, Mr. Li. We know our history. Can we go on to our Biology studies now? We are learning about the systolic system of blood supply in humans.'

Mr. Li's eyebrows sprang up. 'Oh, I see. I will need to get my notes and texts…'

Jonathan's hand was up. 'No, no, Mr. Li. We know what to do. Leave it to us.'

Before he could utter, 'Well, I don't know…,' Mr. Li was watching the whole class immerse itself in reading from thick medical tomes, note-taking, drawing diagrams. Their focus was like nothing he had ever witnessed.

And so their self-learning proceeded with Mr. Li a watchful bystander.

Geography – they all seemed to have travelled the world.

Physics: principles of flight, gravity, acceleration…

Religion: religious icons, prayers, services – all known off by heart.

Soon it was lunchtime. The morning had passed in a flash. This had to be the brightest group of students Mr. Li had ever met. And they were 5AB. What would class 5A be like? A bunch of Einsteins?

A bell rang from somewhere in the darkness. Not the usual school bell but a slow, mournful bell, like a call to prayers. Automatically, the class trooped out without a word. Li, expecting them to be going to the Dining Room, stuffed his papers into his bag and followed them. He was feeling peckish and needed some nourishment – a steak perhaps he laughed to himself. But by the time he was at the door to the classroom, they were gone. Not a sign of twenty teenagers in a matter of seconds. 'Impressive,' Mr. Li's mental note to himself.

Unsure of himself – where he was, where he was to go –

Mr. Li shuffled along the shadowy corridors. The school was silent. Perhaps, he thought, I should find Mr. Singh and report his first impressions but everywhere was deserted. Empty. Silent. He checked his rota for the day and found the lunch break was one hour. One hour to spend. He'd have to eat the shushi he'd bought for tea. Perhaps he could stroll round the grounds except he soon found the external doors were all locked.

'What sort of school is this?' he snorted to himself, retreating from the locked and barred doors.

On the hour, Mr. Li wandered back to his room. Yes, there was the classroom, 5AB Negative. Such an odd title for a class. A blood group for goodness' sake. What a silly coincidence.

Miraculously, the class were all in place, their pale faces turned up to him once more in some sort of expectation. But he wasn't sure he could teach them. They seemed to know everything.

Jonathan stood up. 'In the afternoons, we do reading, Mr. Li. One group is reading the Dead Sea Scrolls. Not in the original of course. (laughter) Another group is working on the Magna Carta and the top group – led by me – is reading Stephanie Myers' 'Twilight Series'.'

Mr. Li frowned. 'Isn't that a bit simple for you?'

In a split second, Jonathan was stood close in front of him, their noses almost touching.

'Mr. Li, do not insult Stephanie Myer. She is a goddess.'

Mr. Li took a step backwards. 'Well, if you insist, Jonathan. I just thought…'

'Don't bother to think, Mr. Li. We'll do the thinking for you.'

And there he was, back amongst the others, opening the pages of 'Twilight Series 8' with a calm smile on his face.

The afternoon proceeded in almost total silence as the students read their various books, occasionally taking notes. They looked at each other from time to time but no words were spoken and yet they seemed to nod or shake their heads in response to a look…almost as if they were…reading each other's minds.

Mr. Li jolted in his seat. If they could read each other's minds – and they seemed bright enough to be able to do that – maybe they were reading his mind as well. Oh dear. He had to be careful now. All those thoughts about the girl in the 7/11 shop. Such a nice smile. No! Stop, you fool!

The class turned as one and smiled at him. Oh my God! The same smile as the girl. How could they have known…?

Mr. Li dived into his bag and pulled out his book of great poems. That would be safe to read.

'Tyger, tyger, burning bright,' intoned Jonathan from somewhere.

Mr. Li looked down at the page. Yes it was there – Blake's poem. 'Tyger, tyger, burning bright…'

Mr. Li pasted a smile on his face but underneath his skin was crawling.

'Yes, good guess, Jonathan.'

'It wasn't a guess, Mr. Li.'

Jonathan's face lowered to his book again.

Mr. Li read the words of the poems for an hour or so but nothing sank in. He was scanning, not reading. He now realised he wanted to be out of this room. Out of this school. He wanted to be somewhere different. Somewhere normal. That wasn't this place. Anywhere. Anywhere away from these…freaks.

Jonathan's face turned upwards and stared at him, accusingly.

'Class ends at three o'clock, Mr. Li. No-one goes till then.'

Mr. Li felt as if cold congee had been poured down his neck. Was that a threat from Jonathan? An order? That was outrageous!

'It's how things are, Mr. Li,' went on the boy, his eyes glassy with indifference.

'Oh, really?' Mr. Li was tiring of Jonathan's superiority. Who was the teacher round here?

'Really.' The word was toneless. Final. A silence fell.

Mr. Li looked at his watch. He'd been wandering around while the class were involved in their reading. No-one seemed to take any notice of him. He was almost irrelevant. Why don't I just walk out? But he knew that wasn't on. He'd signed a contract to be here for a month. And anyway, all the outside doors and gates were locked. Still, if he had to do nothing – apart from not getting annoyed at Jonathan – it was easy money.

'Five minutes left. Can you start putting your books away?'

Still in silence, the class did just that and sat staring at him. Jonathan nodded to one boy who walked to the door.

'You can't go yet. It's not three o'clock.' Mr. Li was imposing himself as teacher.

The boy looked puzzled. 'I was just going to lock the door.'

'Silly boy,' grinned Mr. Li. 'If it's locked, how can we get out?'

The boy clicked the key in the lock nonetheless and carried the key to Jonathan who placed it on the desk in front of him.

Mr. Li swallowed hard. He was going to have to grab the key, unlock the door and march out. Not quite how he'd seen the end of the day. He stood in front of Jonathan and held his hand out.

'Give me the key, Jonathan and we can all go.'

Jonathan's face was an emotionless mask. He was whiter now than he'd been all day. His eyes looked strained with dark rings under them. He almost looks older, much older, thought Mr. Li.

'We're not all going, Mr. Li. That's the point.'

'What are you talking about? What nonsense is this? Give me the key.'

Mr. Li leaned forward, sweeping his hand towards the key but the boy's hand was faster, infinitely faster, and before he had finished grasping at air, Jonathan dangled the key before the teacher's eyes.

'What are you talking about, boy?' Mr. Li's voice was sharp like an icicle.

He wanted to slap that white, insolent face. Hard. Slap away that sneer that hung on the boy's handsome lips. He brought his arm back and his open palm swept down. But before it could move further, Jonathan's hand was on his wrist, holding the arm firm, still.

'I don't think teachers are allowed to do that anymore, Mr. Li.'

'Let go!' bellowed the teacher but his hand was in a vice. He couldn't push or pull against its strength.

'Calm down, Mr. Li. It's all under control.'

Jonathan nodded and Mr. Li was surrounded by the class. Now a boy clamped each arm, each leg. Mr. Li felt dizzy with helplessness. What was going on? He wanted to be out of here. Away from these awful kids but there was no way he could move.

Slowly Jonathan stood up and moved around his desk to stand close to Mr. Li.

'This is how it's meant to be, Mr. Li. Just accept it. Yes, we needed you but not as our new teacher.'

Mr. Li knew his eyes were bulging and rage was building up inside.

'If I'm not your teacher, what the hell am I?'

'Our tea, Mr. Li. Oh, I'm a poet and I don't know it.'

Jonathan's mouth snapped open. Mr. Li stared in horror at the two white fangs that gleamed as incisors. The boy's eyes were now red as he gulped a huge breath of air.

Mr. Li would have screamed had he had the time but in one savage bite, the boy had taken his throat out. He watched, puzzled, through the searing pain as a fountain of blood showered out over the white faces around. He noted, almost with detached interest, how each face licked the blood in pleasure it seemed. Then Jonathan's teeth bit through his sinews, through the neck bone completely and Mr. Li seemed to be falling back into blackness.

Be Careful What You Wish for

Thursday lunch time. Computer Club. Best bit of the week. Well it was for Aziz. Maybe less so for Ali who didn't like being confined in a room, staring at a screen as much as Ali. Aziz was proud of his Computer Club Member card even though he didn't show it because the other kids would call him a nerd. Ali drew rude pictures on his card. But for all their differences, every Thursday they sat together in that mega-low temperature room which only looked out onto concrete walls and they gamed. Oh, how they gamed. And it was one game alone that had them in its spell. James Bond.

Yes, that James Bond. The spy guy, the Brit, the fancy cars, the girls, the chases. The murders, the explosions! They'd played every game there was: Doctor No, From Russia with Love right through to Skyfall and Quantum of Solace.

Aziz always took the Bond role and Ali, the villain…Doctor No, Blofeld…and nearly always Aziz would win the game. Nearly always. He was good and knew he was good. He could even do the Sean Connery voice as they left the computer room.

'Well, Ali…that wash an exshellent game today. You played your besht but it washn't good enough.'

He could roll his 'r's' and make enough shushing noise with his teeth, to be Sean Connery. He couldn't be Daniel

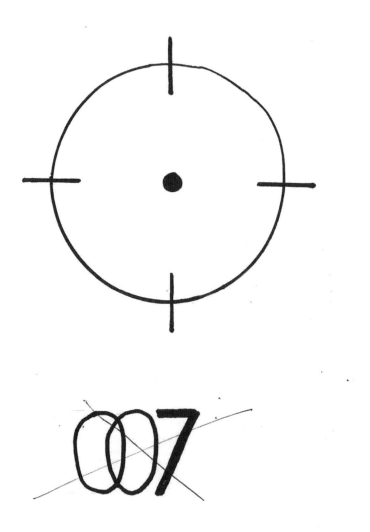

Craig. He just sounded like a fourteen-year-old Asian boy trying to speak posh English. Not Daniel Craig.

Aziz played and gamed James Bond. He had read all the books. He'd seen all the films (some of them when his Mother and Father were out because they didn't approve of James Bond.) 'He always kills us Asians!' his father would shout. 'We're just villains. Stupid stories!'

Sometimes Aziz would try to argue but just when his Bondian logic was working, his father would say, 'And don't contradict me! I am your father!' And that shut James Bond right up. Because if he went on, his father would go berserk. And Aziz didn't want that.

It was coming up to the Chinese New Year. Aziz and Ali, by this time, had played all the Bond games at least once and sometimes three times. 'Goldfinger' was a favourite. Ali joked that the air conditioning in the room meant it had to be called 'Coldfinger'. Aziz had beaten Ali again and for once Ali didn't laugh when Aziz did his awful Connery voice. Ali didn't speak much after that day. Perhaps he was a bit sick or something. That evening, after dinner, Aziz was in his little cramped room. He'd finished his homework and opened up his email. He even tried the Connery voice himself but it wasn't very good.

'Well, Mish Moneypenny. What do we have here?'

Across the email was stamped 'Security Classified'. Aziz knitted his brow and read the script.

'Aziz Kandul,

Your on-line interest in espionage and counter-espionage has been identified by our department. Your personal profile, which we have checked carefully, reveals that you are a number one lead candidate for our cadet espionage team. The Government is actively seeking to recruit like-minded young individuals who might serve their country in particular ways. All this, of course, must remain wholly confidential. None of this email must be forwarded or shared. We are closely monitoring how it is dealt with by you. Should you do anything other than what we tell you, you will be dealt with accordingly as a traitor.'

Aziz glanced both ways and behind him. Were they watching him? Of course not. They were tapping his internet line. His eyes re-focussed on the email.

'If you agree to contract into our elite cadre of cadet espionage agents, you will be well rewarded. Details of these rewards will be sent to you should you agree to this invitation. Just send a reply 'Agree' by return. Remember – we know everything you do.'

No name. No company or department logo . A plain script. Could it be true? He had no way of testing it. If he asked someone, 'they' (whoever 'they' were) would know and he would be punished. He decided he would think about it and would respond the next night after school.

It was a quieter, more subdued Aziz who went to school the

next day. Ali seemed to be noisier than usual, laughing more. Was he even noticing that Aziz was…preoccupied? How he wanted to tell Ali what he had been offered! But he couldn't because…they were watching him, maybe?

At lunchtime he stayed in the home room, looking out beyond the school gates. Was that black car parked there this morning?

'Come on, Aziz. Let's go play some hoops,' shouted Ali, running out with two other boys. But Aziz shook his head. He didn't like basketball anyway. A stupid game. He had more important things on his mind. To become a spy or not. What did that mean? What sort of spy? What would he be expected to do? Kill people? Surely not. Oh, it was easy in a game. Point the gun. Press the button. Kaboom! Ah, but in reality? He squeezed his eyes tight shut to stop the visions till his ears seemed to whistle.

'You are not leaving the table, Aziz, until you have finished your dinner.'

His father's voice was heavy and dark.

'I'm sorry, Papa, but I'm not hungry.'

'He didn't eat his packed lunch,' intervened his mother, very unhelpfully he thought.

'It's okay. I'm okay. I need to be careful what I eat. We did it in Food Technology. 'Try a few days of eating less and see how different you feel'. That's what the teacher suggested so I'm trying it. It's no big thing. I'm just doing an experiment. On myself. It will be over in a few days.'

Aziz impressed himself with his own ingenuity. He had almost convinced himself with his argument. Perhaps he was cut out to be a spy.

'Well, it's a waste of good food. Pass his plate, Mother. I'll finish it.'

Aziz looked from one frowning parental face to another. They had accepted his story.

'I have extra homework tonight. May I leave the table?'

Another lie. The two heads nodded solemnly and two pairs of eyes followed him. He knew he must act normally. Just be natural. He hummed the James Bond theme as he left the room.

Aziz's finger trembled as it hung over the 'Send' button. One simple, downward movement would seal his fate – whatever that might be. Life could never be the same again. His eyes were locked on his name, there on the screen. 'Aziz Kandul'. The words morphed into 'James Bond' and back to 'Aziz Kandul'. What had he got to lose? Ali would be so jealous of him just as Aziz was jealous of Ali's skill at basketball. At cricket. At athletics. Now it was his turn to be top dog – not in a fantasy game but in real life! He hit the button.

Almost immediately a new message appeared.

'Aziz Kandul 00145

Please note that in future you will only be known by your code number. You are now part of the government Junior Espionage Cadre. Full details of your contract can be found on our official website. Use only your personal number to access.

Your first task is as follows:

Tomorrow, April 16th, you are to meet another cadet agent at the Café club Coffee House in Pacific Place at 5.30 pm. You will receive an envelope which you must not open on any account until you are told to do so by a senior officer. The other agent's name is Melanie. When you have the envelope safely, send a blank reply to this email. Good luck.'

Melanie. A girl. Perhaps someone who is like him. Or perhaps older? A beautiful seventeen-year-old who would hold his hand and smile as she passed…

'Have you started your homework yet?' You're still on that wretched computer!'

His father was in the doorway glaring at him. Aziz slammed the lap-top shut.

'Just starting, father. I was checking the details on the school web site. We all do it.'

'Pah!'

His father turned angrily on his heel and disappeared. So this was it. His first assignment. Quite easy. Meet the beautiful girl. Get the letter. Check back in. Simple!

Ali kept looking at him the next day. Aziz would catch him out the corner of his eye then Ali would look away.

'What are you looking at, man? Do I have a snot bubble or something?'

'No. Nothing. You look different that's all.'

Different? More grown up? More serious? Better try to look and sound normal. Ali was looking at his head.

'Aziz. You've combed your hair different. You joining a boy band or something?'

It was true. Aziz had taken some of his mother's hair oil and had rubbed into his mop. He thought it had looked good. Better change the subject.

'No, Ali. I'm fine. Fancy some hoops after school?'

'You don't like hoops.'

'I like spaghetti hoops. I do like basketball. Just today. Just ten minutes.'

Ali was looking at him curiously. He calculated. 3.45 latest. Bus to Pacific Place: twenty minutes. Then…Melanie. How he wanted to tell Ali everything. The invitation. The girl. The secret message. Ali would be so in awe of him. What a day this was turning out to be.

'You're weird, Aziz. Do you know that?'

The bus – a damned red top minibus – seemed to be totally stuck in the Mid-Harbour Tunnel traffic. Cars, shining with rain, stretched in every direction. Taxi drivers leaned on their horns. There was a symphony of parping. The bus driver kept slapping the steering wheel. Passengers stared glumly out, not speaking. But Aziz couldn't sit still. He shuffled his feet. He leaned back. He leaned forward. He checked his watch. Five o'clock. They were only just entering the tunnel. He'd never get there on time. Melanie would slowly rise and, looking round for the last time, would leave. Aziz had failed his first assignment. Probably his last assignment. If only he'd come away from the basketball ten minutes earlier. But Ali had insisted he stay and play out the game. He knew Aziz had to go (a dental appointment – another lie) but kept passing the ball to him. Aaaaagh!

It was 5.35 when, red-faced, sticky with sweat, his shoulder bag cutting into his arm, Aziz reached the Café Club, banging his head on the glass door when he pushed rather than pulled. The loud crack of his skull on glass turned heads. The café was full! How would he know who Melanie was? He couldn't go up to each girl and ask them: 'Are you a member of the Espionage Cadre? Do you have a letter for me?'

His eyes flicked from one girl to another. The one in the corner with the orange fringe? She was talking, foreheads close, to a boy. Probably not. The chubby-faced girl looking down at her cell phone? She didn't look the sort. Too… homely. She wasn't looking up for anyone either. What about the tall girl to his left? She was reading a book. She was much more likely. Her glasses made her look intelligent. And spies had to look intelligent. Aziz automatically ran his fingers through his hair trying to make it look cool. Intelligent hair.

Yes – that would be Melanie. He was sure. He stepped across to her and stood awkwardly in front of her.

'Melanie?' She slowly raised her eyes from the book.

'No. Go away, you stupid little jerk.'

He turned and shuffled to the counter. If he got a coffee, he could survey the scene more…naturally. He ordered a flat white and when it came, he slowly stirred the cardboard pot, looking round at the tables and the faces that hovered above. There must be twenty girls here. How could he possibly find out which one? Shout her name? Shout his own name? Shout out, 'Who's the spy here?' No. No. None of this would work. He had to admit, he wasn't very good at all this. This spy stuff. Lying to his parents? Yes. Spy stuff? No. It was good in his head – easy, logical, do this, then this. But in reality? It was soooo difficult.

That was when there was a gentle tap on his shoulder.

'Aziz? I mean, 00145?' a soft voice seemed to murmur.

He whirled around. Coffee slopped out of his cup onto the table. He stood up. Before him was a young girl – maybe two years older than him – as tall as him (she looked him straight in the eyes). She was in casual gear – a puffa jacket, shiny silver, tight jeans with slits in the knees. A silver hair band kept the curtain of black hair from her face. And she was…pretty. Very pretty. Aziz's heart jumped about in his chest. He realised he was staring and he'd said nothing.

'Aziz?' she repeated.

What should he say? 'Aziz Kundal'? Should he give his

full name? Perhaps 'Kundal. Aziz Kundal' in a James Bond voice. Bond always said that. 'Bond. James Bond.' It sounded smooth. Assured. Or should he just say 'yes'? She was looking at him, waiting. She put her head on one side and smiled to encourage some sort of response.

'Aziz? Aziz? Yes, that's me.'

'I'm Melanie. We're supposed to use code numbers only but it sounds a bit weird here.'

'Yes. Melanie. Yes, I knew that. Not who you were but that your name was…is…Melanie. I was told by…'

Her raised palm stopped him.

'Aziz. We are here today only for me to pass you this letter.'

She held up a brown envelope. He could see his number written on it. Aziz reached out and took it carefully between his fingers but she didn't release it.

'Tomorrow, we can meet again here and it will be different. It will be because I want to meet you properly. Until then…'

She leaned in and kissed him lightly on the cheek. He felt his cheeks burn red under their brownness.

'Right. Yes. Thank you. Tomorrow. Melanie. Aziz. That's me. Meet.'

The top of his head seemed to be lifting off and all the words in there were scrambling around trying to get out. He opened his mouth to let a few escape but she was gone. Through the crowd, out the door (without banging her head), down the escalator to the MTR. Should he run after her? Ask her more questions? Too late. He was rooted to the spot with red cheeks and a letter flapping feebly in his hand. Cold coffee was dripping off the table onto his shoes. He looked at the envelope. He knew he couldn't read it here in public. He would have to wait. He folded it carefully and put it inside his jacket – the inside pocket. Safer.

As he meandered home, Melanie filled his head. Her face, her hair, her puffa jacket, her jeans. What was she wearing on her feet? Oh god. Spies were supposed to be observant. See details. Remember stuff. He couldn't remember what clothes she was wearing. He could remember her face. The single lid eyes that seemed almost too far apart. The snub nose. The rose pink…

'Look where you're going!'

Aziz had nearly bumped into a big, thick-set white man in a black suit. The man had grabbed him and guided him to one side. Aziz was suddenly alert. He stared at the grey stubble haired head of the man walking on now. Then his espionage mind clicked into gear. He slipped his hand inside his blazer. The man had been trying to pick-pocket the letter. It was part of the test. His fingers rested on the envelope. The man had failed. Aziz's role was still intact. This was what he should expect from now on. Everyone was a potential enemy. He had to be on his guard. He mustn't relax for a moment.

Dinner was the usual thing. Father complaining about work. The people. The office. The people he had to phone.

Mother complained about her work. The people. The office. The people who phoned her. Aziz told them he had had a 'nice day' at school. That seemed to keep them from asking for details.

His jacket was hung on the back of his chair so every now and again he would feel to check for the letter. Each time he felt it there he rewarded himself with a big bite of beef noodles.

'Stop fidgeting with your blazer, Aziz. You'll wear it out.' Snapped his father as his fingers paddled inside the pocket again.

'I like my blazer.'

Why had he said that? It was a dumb thing to say. They knew he didn't like it. He'd told them often enough.

'Well. That makes a change,' joined in his mother, acting as reinforcements to the cavalry charge of his father. 'Perhaps you'll hang it up properly then.'

Aziz rose and reached around for this article of such family interest.

'Not now! Sit down and finish your meal!' blazed his father.

Slowly, Aziz lowered himself onto his chair. This wasn't going so well.

How does time stand still? It had now. The meal was going on and on and on. Just when he thought it was all over, his mother would rise with a smile on her face and produce another dish of food. It was no good arguing. That would slow things down. If he didn't eat, they would interrogate him. That's more time. All he wanted was to get into his room and check in with the espionage group. He couldn't just open the letter, could he? He had to have the authority to open it. The direction. He shovelled in more food. By the time the meal was ended, he was feeling bloated like a balloon and he wanted to be sick. But that might have been nervousness. What might the letter tell him to do?

Aziz slumped in his seat, his eyes closed, a cold sweat on his brow and upper lip. He held the opened letter limply in his fingers. He couldn't bring himself to look at it again. To read those words again. He had checked his e-mails and there had been the simple message:

'Open the letter. Do exactly what it says.'

And what had it said? He felt the top of his head lifting. Now he really felt sick. The milk pudding he'd consumed was rising in his throat. There was only one thing to do.

He rushed to the toilet and his guts roared like an angry animal. The beef and rice and dumplings and fish gushed out of him in a yellow, barking stream, splashing up all around the lavatory seat.

There was an urgent knock on the bathroom door.

'Aziz! Aziz! Are you alright? Are you being sick?'

A long, sticky string of goo hung down from his nostrils. He wiped it away with some toilet paper.

'I'm okay, Ma. I just ate too quickly. I'm fine.'

'Open the door and let me check.'

Check what? He thought. Me or whether I'd vomited all over the bathroom? He flicked the lock and his mother bustled in, grabbed a towel and rubbed his face. She looked sourly into the toilet bowl and flushed the mess away.

'You shouldn't eat so fast, Aziz. I warned you. You are like a hungry dog sometimes.'

Thanks for that, Mother, he thought.

'You look dreadful. Go to bed and get some sleep. Homework can wait. You are a sick boy.'

He still felt sick in a different way as he shuffled back to his room. The letter lay there, gloating at him. His hand still trembling, he read it one more time before tearing it into tiny pieces. He would have to do what it said. He would just have to. However much he didn't want to do it, he had to. It's what he had signed up to and there was no going back. That would be cowardly. Sleep did not visit Aziz that night.

'You look terrible!' laughed Ali the next day when they met. Aziz didn't smile. Couldn't smile.

'What happened?' His friend's arm was round his neck, squeezing in a friendly way.

'Oh – just sick. Vomit stuff, you know. Too much of mother's rich food last night. I'll be fine.'

But he wasn't fine. He was white and shaky and seemed

angry with Ali whenever he spoke.

'What's the matter with you, Aziz? I've not done anything wrong, have I?'

Aziz's heart sank.

'No. Of course not. You are my best friend.'

'Then let's be best friends, like normal.'

Aziz forced a smile onto his lips.

'Okay then, Ali. Let's skip school this afternoon. I've got a good excuse if we get caught. It's only Chinese History class. Wong will never know we're not there.'

Ali looked Aziz carefully in the face.

'You mean, YOU want to skip school. Well, well. Look, man. You're not well. Sure. Let's do it.'

Ali smiled broadly again. Aziz hooked his bag over his shoulder as they sidled off as casually as possible towards the back gates of the school.

Ali was sprawled across the kitchen table slurping at a Cola Aziz had carefully poured him. Why did Aziz's friend smile and laugh so much? Aziz, for once, found it hard to talk to him where usually their conversation flowed like a river of jokes and comments and nonsense. Even when Ali began his James Bond impressions, Aziz could barely smile.

'That impression is terrible,' said Aziz.

'You liked it the other day,' chirped Ali.

'That was the other day. Today is today. It was terrible.'

Ali took another slurp of Cola and burped loudly.

'Shall I do Daniel Craig then?'

'No. Don't even try to do Daniel Craig. It will be awful.'

'Or even awfuller?'

'Probably.' Aziz dropped his eyes, not wanting to catch Ali's smiling look.

There was a silence. Aziz fiddled with the kitchen drawer, opening it and looking down as if there some Egyptian treasure hidden in the drawer. He was sworn to secrecy to the espionage group and he didn't want to lose this great chance to be…someone. But he had to tell Ali some of it – surely? Especially now he had his special instruction.

The silence thickened in the air. Ali put his head to one side, staring at Aziz.

'Something wrong, Aziz? You still feeling rotten?'

Aziz shook his head. 'No. Not ill. There's something I should tell you. Something really important. But I'm not supposed to tell you or talk about it.'

Ali sniggered then forced himself to stop, pulling his face to serious.

'Then maybe you shouldn't tell me, if that's the case.'

Aziz bit his lip hard. He tasted sour and bitter blood.

'I think I have to. In the circumstances. So you understand.'

Ali smiled. 'Understand? Sounds very…mysterious.'

'It is.'

Ali almost whispered these words as if they were almost impossible to get out. He clutched at the kitchen drawer, still looking down.

'I don't know why it has happened, all this. But it has.'

Ali was looking at Aziz over the rim of the cola cup, watching him it seemed with glittering eyes. He was amused by his friend's troubles.

'I've become a member of a government espionage group.'

Ali snorted into his cup, making bubbles blow up into his face.

'Oh yes, of course you have. An espionage group. Aziz is now a spy, are you?'

'It's true, Ali. On line. A cadre of young spies. And I had to meet this girl, Melanie.'

'Oooh, like a dating agency. Was she nice?'

'No. Not like dating agency. It was for real. Serious.'

'Don't tell me, Aziz. You fell in love? Is that what you wanted to tell me?'

'No. No. Nothing like that.'

'It's okay to fall in love at fourteen. I have, dozens of times. Was this girl, Melanie, nice looking?'

'Yes. But that's not the point. It's not about falling in love. That's not the point. Not the point at all!'

Aziz crashed his hand down on the kitchen drawer, making the utensils inside rattle.

'Well, get to the point, Aziz.'

Aziz glanced at Ali then looked down again.

'To prove my worth, my…ability…my commitment…I was given a task.'

'A task,' grinned Ali. 'Was it to do with Melanie?'

'No, Ali. It was to do with you.'

'And how did they know about me?'

'They didn't. But I did.'

'This is all sounding a bit weird, Aziz. So what is the task?'

Aziz swallowed hard then surprised himself at how quickly he acted. The kitchen knife was in his hand, a long blade six inches long, used for vegetables. His mother kept them very sharp. He stepped across to the table and, while Ali was putting his head back, draining his coffee, he drove the knife into Ali's ribs, just below the ribs. The blade went in easily, through the boy's shirt, his skin, his tissue, his…

Ali screamed, the cola spraying across his face. He fell backwards over the chair, the handle of the knife sticking out at a strange angle. Aziz was rooted to the ground, watching Ali writhing in pain, as if this was some odd act that was unconnected to what he had just done. He stared at his empty hand.

'What are you doing, Aziz?' gasped Ali, clutching at the knife and the widening dark stain on his t-shirt. 'Help me, for God's sake!'

'I had to do it, Ali. I was…under orders from the espionage group.'

Ali let out a deep grunt of pain. 'The espionage group?' Ali's face squeezed with pain. Fury burned in his eyes. 'There is no espionage group.'

Aziz now looked desperately into his friend's eyes, shining

with pain and anger. What nonsense was this?

'What do you mean? There was no espionage group? It was all on-line. I joined.'

Ali gulped for air.

'It was a joke, you stupid. I set it up as a prank.'

'You did what?' Aziz's mouth hung open.

'I did it all on-line. It was a joke. The espionage group. Everything.'

'What about Melanie?'

'She's a friend of my sister. I roped her in. She thought it would be funny too.'

Aziz clutched his head and shouted up at the ceiling, frantic.

'Oh God, no. Ali! Why did you do it? Do you hate me so much?'

'Aziz. Get a doctor. Call the hospital. This is really hurting. I could die. Aziz!'

Aziz reached down and took hold of Ali's t-shirt.

'Not until you tell me why you did it. Why did you want to humiliate me? Friends don't do that to each other.'

'I never thought you'd believe it. I didn't think you were that dumb. Now call the hospital! I'm bleeding badly.'

Aziz straightened and walked across to the window and stared out across the estate – the concrete blocks reaching up to the sky, the clothes hanging from the windows, the empty playground way below, one old man on an exercise machine pedalling furiously, going nowhere. Tears filled the boy's eyes and everything became blurry.

'What would James Bond do in this situation, Ali?'

Ali extended a bloodied hand towards his friend. 'Call the hospital!' His voice was breaking with the pain.

It was Bond's voice that replied, in the rich, warm Scottish tones of Sean Connery. Aziz had never been able to do that voice till now and it was perfect except that Ali didn't care a damn.

'Well, Mish Moneypenny. Thish ish a mesh. I think itsh time to make an eshcape.'

Aziz reached out and opened the window.

The story is based on real events in the UK when one boy was tricked on-line by his friend and believed he had to kill his friend (which he tried to do by stabbing him). The friend survived.

Truth is stranger than fiction quite often.

Notes:

1) 'James Bond': a hero figure created by the writer, Ian Fleming, who has featured in dozens of famous and very popular films from 1963 to the present.

2) Sean Connery: a Scottish actor who played Bond in the first Bond movies.

3) Daniel Craig is the actor who currently plays Bond (2019).

Auto Message

Telephone in apartment rings. Girl, 15, answers.

'Hello. Nei ho.'

'This is the Sacred Heart Secondary School. Am I speaking to the parents of Anna Tsui? Please answer yes or no.'

'This is Anna Tsui.'

'Please answer yes or no. Am I speaking to the parents of Anna Tsui?'

'Er…no.'

'Am I speaking to Anna Tsui?'

'Yes, I am Anna Tsui.'

'Anna Tsui. This is the third day of absence from school. Please answer yes or no.'

'Well, strictly that was a statement not a question.'

'Please answer yes or no.'

'Yes…yes, this is my third day.'

'Anna Tsui. Do you have a doctor's note to cover your absence? Please answer yes or no.'

'No. No, I don't. I haven't been to…'

'Anna Tsui. Have you visited your doctor? Answer yes or no.'

'No. No, I haven't been to the doctor. It's a waste of time. He never bothers to…'

'It is a school rule that any absence from school requires a medical certificate. Anna Tsui, do you intend to obtain a certificate? Please answer yes or no.'

'I don't because I think it's a complete waste of…'

'Please answer yes or no.'

'No. No, I'm not.'

'Anna Tsui, do you know that if this rule is broken you could be fined, taken to court or even locked up in prison? Answer yes or no.'

'What? But I've only been away three days. I've been…I've been ill.'

'Please answer yes or no.'

'No. No. No. I didn't know I could be put in prison for missing school. What kind of punishment…'

'In accordance with the school rules, you must be prepared to surrender yourself to the school authorities. Do you understand? Answer yes or no.'

'Prepared to surrender? I've been ill not causing a riot or robbing a bank.'

'Please answer yes or no.'

'No. No. I mean, yes. I do understand that. But this is crazy. Is this a joke? Is that you, Sara? Have you set me up?'

'You have thirty minutes to surrender to representatives of the school authority. Do you understand? Answer yes or no.'

'Oh shut up. Sara. I know it's you.'

'Please answer yes or no.'

'You're being really annoying, Sara. The joke is over. I get it.'

'Please answer yes or no.'

'Yes, Sara. I understand. Are you coming to arrest me?'

'Our representatives have the authority to enter your home. Do you understand? Answer yes or no.'

'Enter my home? Like police? Or a swat squad? Don't be so stupid.'

'Answer yes or no.'

'Yes. But I'm not letting them in. They have no right. Oh, this is soooo annoying. Stop it!'

'Our representatives will be with you in three minutes. Do you understand? Answer yes or no.'

'Three minutes? How can they…going to be here? I'm out of here.'

'Answer yes or no.'

'Yes, you moron. I understand. But I'm not staying here to be arrested for no reason.'

'If you try to leave, you will find that the area will be closed down. You cannot run away. Do you understand? Answer yes or no.'

'Yes, you idiot! Of course I understand but you must be mad if you think I'm going to stay here and…Who's there? Who are you? Stop knocking! Stop knocking! I'm here. Who the hell are you? What do you think you're doing? You're busting the door! Stop it! Stop it! Leave me alone! Get out! You have no right! Don't touch me! Don't touch me! Don't come any closer or I'll…'

'Representative. Have you concluded the necessary arrest? Answer yes or no.'

'No.'

'Did the subject evade arrest? Answer yes or no.'

'Yes. She jumped.'

'Has another school problem been permanently solved? Answer yes or no.'

'Yes.'

End of Term

This 'playlet' was produced after some drama improvisation workshops in a Hong Kong secondary school.

'End of Term' might be considered too dark, too violent. It is more realistic than most of the other stories as the basic situation is ordinary and every-day-a student visits a teacher in their office after school. What follows is not what happens in schools generally. However if what happens makes the story 'unacceptable' then how do we think about the dozens of school shootings in the U.S. every year where students and teachers die in bloody circumstances? In the UK it is not unknown for a teacher to be murdered in front of their class. Across the world, teachers are variously attacked by their students. So, whilst the story may not represent what happens in Hong Kong schools, it does raise issues about violence in schools. The story most certainly does not endorse or encourage violence to solve problems, as the outcome would suggest. Hopefully it will generate important debate among readers about the use of violence.

Scene: A TEACHER'S OFFICE. The clock on the wall says 5.30 pm. The calendar date says 21 July. The teacher, MISS KWAN, is sitting at her desk, drinking coffee and writing notes. She stops writing, smiles, closes the folder and places it on top of a small pile of other folders. She leans back, puts her hands behind her head and closes her eyes.

KWAN: (to herself) The last one. The last report. The last day of the year. Thank God.

(She drains her coffee cup and starts to collect her things together, putting them into her handbag: pens, spectacles and mobile phone. She tries to activate it but it is clearly dead. She drops it into the bag. The office door rattles. KWAN looks at it, puzzled.)

KWAN: Hello? Anyone there?

(The door swings open and ALVIN, a sixteen-year-old student enters. He is in school uniform but is untidy – shirt hanging out, school jacket crumpled, school tie looped round his head.)

KWAN: (hesitant) Alvin! What are you doing here? School closed two hours ago. Thought you'd be off for the summer holiday, not hanging around here.

ALVIN: Um…errr…

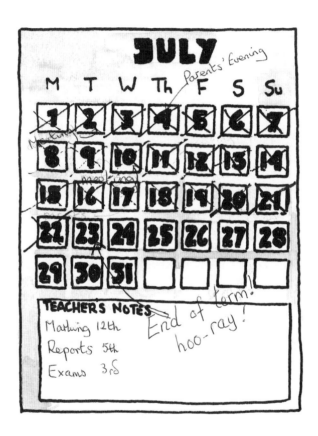

KWAN: So what can I do for you, Alvin? I don't have much time – I'm just about to go.

ALVIN: No, you can't do that. You have to stay.

KWAN: Well, no, Alvin. I do have to go.

(KWAN stands and begins putting things into her bag.)

ALVIN: No, you stay.

KWAN: Well, Alvin, that's a bit rude. I am…was your teacher. You shouldn't speak to me like that.

ALVIN: That's it. You were my teacher. So now I can speak to you how I like.

KWAN: Look, I haven't got time for this. If you've come just to insult me…

ALVIN: No. Not just to insult you. I've got questions.

KWAN: About your exams? You'll have to wait for the results. Perhaps come in and see me when you get the results.

ALVIN: No exams. Not results. Other questions.

KWAN: Other questions? What would they be then?

(She looks at her watch.)

ALVIN: It'll take time, Miss Kwan.

KWAN: Yes, but I don't have…

(KWAN goes to walk from behind the desk but ALVIN moves quickly forward blocking her way.)

ALVIN: Sit down!

(KWAN looks at him fiercely. He is head and shoulders taller than her, solidly built, an athlete. She slowly sits down.)

KWAN: Alright, Alvin. Now what are these questions you have for me that are so urgent? Why don't you sit down?

ALVIN: No. I'll stand.

KWAN: You'll stand. Okay. Well?

ALVIN: First. Why did you hate me?

KWAN: What, Alvin? What makes you think I hated you?

ALVIN: You did. You hated everything about me. My work. My way of speaking. My…clothes. My hair. Everything.

KWAN: I didn't hate you, Alvin. It's just that…all those things were not…what we wanted at this school. You didn't do any work. You were never in proper uniform. Look at you now.

ALVIN: I don't have to wear this stuff anymore. It doesn't count anymore.

(ALVIN tears the tie from his head and throws it on the desk. He tears his shirt front open.

KWAN: (Moving back in her chair) No, it doesn't matter anymore. But when you were a student, we expected…I expected you to keep to the rules. They were for everyone.

ALVIN:: I couldn't do the work. You said so. You said I was stupid.

KWAN: I'm sure I didn't. I would never have said…

ALVIN: You did. You said it a lot. 'That's a stupid thing to say.' 'That's a stupid idea.' I remember. I remember every time you called me stupid.

KWAN: What I said was, your words were stupid, or the

idea was stupid. Not you. That's different.

ALVIN: No, it isn't. If the words or ideas are from me, then they are me. So I must be stupid.

KWAN: Look, I'm sorry, Alvin, if I gave you the impression that…I really do have to go.

(KWAN rises and picks up her bag.)

ALVIN: Sit down! I haven't finished.

(ALVIN pushes her back into the chair that she nearly topples over backwards.)

KWAN: Alvin! There's no need for that.

ALVIN: Oh yes there is. You're not just walking out on me.

KWAN: You can't do this, Alvin. You can't force me to stay here.

ALVIN: Just try and leave then. I think you'll find I'm right.

KWAN: I answered your questions. Now let me go. I have to meet someone.

ALVIN: You're meeting me at the moment. You're going

to be delayed.

KWAN: Look, Alvin. What is it you want? I've got some money here…

(She rustles in her bag and pulls out a couple of one hundred dollar notes.)

Take it. Go on. Spend it on whatever you like. A sort of end-of-term present from your teacher.

(ALVIN takes the notes and throws them in her face.)

ALVIN: Think you can just buy me off, eh? I want justice. For what you did to me.

KWAN: Justice? Justice? You don't understand the word, Alvin. You mean revenge.

ALVIN: There you go again. Saying I'm stupid again.

KWAN: I'm not saying that. I'm saying…

ALVIN: I'm tired of what you're saying. I've had to listen to it for years. Saying this. Saying that. On and on. Do this. Do that. Don't do this. Don't do that.

On and on.

KWAN: It's what we have to do. We're teachers. It's for your own good.

ALVIN: But it hasn't done me any good, has it? You said I couldn't do exams. A lot of good you did me. I've got nothing now. Nothing. Been in school since I was five and I've got nothing. Nothing.

KWAN: You have. You've learned all sorts of things.

ALVIN: And they're worth nothing. I've got nothing. And it's your fault.

KWAN: Well, that's not fair, is it? Blaming me for everything at school. What about others? If you're going to blame anyone, blame the Headteacher. Blame them all. What's the difference?

ALVIN: The difference is, Miss Kwan, you're here and they're not.

(ALVIN leans on the desk looking down at KWAN.)

KWAN: This has gone far enough. If you don't leave, I will call the police.

(KWAN reaches for her bag but ALVIN snatches it away before she can get it. He takes out the mobile, presses the button.)

ALVIN: No battery left. No calling the police then, eh Miss. No calling anyone.

KWAN: Just give me that back and I'll be on my way. I won't say anything to anyone about this. Alvin, I'm sorry you feel this way. But I have to go.

(KWAN takes the bag back from ALVIN and rises from her chair, looking him firmly in the eye.)

KWAN: I wish I could help you, Alvin. But I can't. Now please go. I have to lock up.

ALVIN: (suddenly very angry) Sit down, Kwan! You're going nowhere.

(ALVIN draws a kitchen knife from his belt, till now hidden under his jacket, and stabs it into the desk. It quivers between them. KWAN stares at it in horror and slowly sits.)

ALVIN: You are not going anywhere. You are going to explain to me why you've ruined my life.

KWAN: Who says your life is ruined? You're just sixteen. There's lots you can do.

ALVIN: What? With no exams? You told us without exams we'd be getting nowhere. Well, here we are. Nowhere. And you're here too.

KWAN: It's not 'nowhere'. Life's not just about taking exams.

ALVIN: Then you were lying when you said we had to get them.

KWAN: Not lying, no. I was trying to describe…trying to get you all to work hard. Do your best.

ALVIN: Except I didn't get that far, did I? You closed the door on me.

KWAN: You hadn't done any work for two years! How could I let you do exams? You would have failed them all. What would the other students think? I couldn't just let you…

ALVIN: You shut the door.

KWAN: I didn't just 'shut the door' as you keep saying. I had to use my judgement. My professional judgement. As a teacher.

ALVIN: And now I'm using my judgement. Cos now I'm the judge. And the jury. See? I do know about justice.

KWAN: Alright, Alvin. So you think you have justice. Now I have to go and so do you. Enough of this nonsense. It's getting us nowhere.

(ALVIN explodes and leans aggressively over her.)

ALVIN: This isn't nonsense! It's my life you're talking about. The life I won't have.

KWAN: That was your choice. You chose to do nothing. You chose to fail. It's your right to.

ALVIN: I wanted help and you didn't give it.

KWAN: You've got a strange way of asking for it. Coming here with a knife seems a bad way of getting help. How can I help you when you wave a knife at me?

ALVIN: You wouldn't listen if I didn't. You wouldn't do anything.

KWAN: It's all too late. Too late to do anything. I can't give you back your time and say 'Have another go.' It is over. As far as I am concerned, it's over. You have to find your own way now. A different way. You didn't want my way. Our way, here at school.

(KWAN herself is angry, rising to confront ALVIN.)

KWAN: So just get out of here. Leave me alone. Go back to your life and leave me in mine. I'll look after this.

(KWAN plucks the knife from the desk and holds it in front of her.)

KWAN: Go on. Just go. Leave me alone. I'm sorry you failed. I'm sorry you feel bad about everything but there's nothing I can do anymore. You're not a student here anymore. It's all over.

ALVIN: Give me the knife.

KWAN: Just go, Alvin and I won't say anything. I won't report you. Leave now for your sake.

ALVIN: Give me the knife.

KWAN: You don't need it. It just makes more trouble for you. Now, go. Please go.

ALVIN: That's my knife.

KWAN: Not at the moment, it isn't. Now leave the room.

(KWAN reaches in her bag and pulls out her mobile phone. She presses the buttons, knowing the phone is not working.)

KWAN: Police, Alvin. As a teacher, I have an alarm number. They will be here in minutes. Why don't you just run? Go and I won't report you.

ALVIN: I don't trust you. I don't believe you. That's my life.

(ALVIN makes a dash at KWAN who is holding the knife in both hands. He stumbles into the blade and it drives up into him. KWAN falls backwards with ALVIN shouting in pain on top of her. His hand closes round her throat. She tries to break his grip but can't. Her screams choke in her throat as her breath fails. Her face darkens as his grip stays. He bangs her head on the floor, clinging onto her neck. Quickly she goes slack beneath him. He lets go, tries to rise but falls on her again. He is bleeding heavily. He howls in pain then collapses. Both lie still.

Outside the sound of birds singing. A plane passes overhead. The telephone on Miss Kwan's desk rings.

Home for Dinner

Amie stared at her friends, her eyes wide. Were they lying? Were they joking? Were they just trying to shock her?

'It's true, Amie,' Zi said, looking her in the eyes. 'That's what happened.'

Dani grabbed Amie's shoulder. 'It's in the newspapers. It's been on the TV news. On Facebook. I've seen it. It's happening now.'

Amie shook her head. 'People eating each other? Oh, come on. It's too gross.'

'Yes, it is,' said Zi, 'but that doesn't make it untrue.'

Amie tried to stand but two pairs of hands pulled her down.

'And it's happening in this city, now.' Dani's mouth was near Amie's ear as she whispered it.

Amie looked around. 'Why isn't anyone doing anything, then? Where are the police? The Army?'

Zi shook her head. 'The police. The government. Everyone is denying it. Saying it is 'fake news' – just made up stories.'

'Not very funny, that's for sure,' said Amie. 'The thing is, what are we going to do? How many of these...what are they?'

'Ghouls.'

'How many? Could be one or two. Could be dozens.'

'Hundreds,' added Dani.

'Thousands,' elaborated Zi.

Amie pulled away from them and stood looking out of the classroom window across the city. She jabbed her finger at the city.

'So, out there, waiting for us are…'

'Ghouls,' chorused Zi and Dani.

'So how do we get home? How do we know who is a ghoul and who isn't?,' asked Amie, spreading her arms.

'You don't know. They look the same. They look normal.'

'What,' said Amie, 'normally eating humans? That sort of normal.'

'Amie,' Zi's tone was like a teacher,' they don't go around yelling, 'Hey, everyone. I'm a ghoul!' No, they're just normal. They wait their time. That's what the guy says on the TV.'

Amie turned to look at her friends.

'So what are you guys going to do?'

Zi and Danni looked at each other, bland and almost bored. Then Zi spoke.

'We're gonna wait till it's dark. Then get home as fast as possible. If there are ghouls around, they're less likely to see us.'

Amie looked at her watch. 4.30 pm. It would be two hours before it started to get dark. Could she wait that long?

'I'm not waiting that long. If it's dark, they can't see you maybe but you can't see them. And they might have multiplied by then. I'm going now. You can wait here if you like.'

Dani and Zi shook their heads.

'Sure. We understand. It's a risk either way.'

Amie packed her rucksack quickly: maths books, exercise books, her reading book, Bram Stoker's 'Dracula' crammed on the top. She turned the front cover showing a vampire biting a victim away and pushed the book deeper. Just a story. Another fake. She stood upright, pulled the bag onto her back and faced her two friends.

'I'm going now.' Tears welled in her eyes.

'Will we see each other tomorrow? Will there be any school?'

The two looked at her blankly. 'We don't know. We don't know anything anymore.'

Amie held her arms out. 'Hugs for friends, forever.'

The three embraced.

'Hugs for friends forever,' chimed Dani and Zi.

Or HUGZ4FRNZ4EV as they put in their texts to each other.

'Text me when you get home,' smiled Zi.

'Sure. And you two. Do the same. OK?'

'Okay. Now GO!'

Amie stepped away from them, brushed the tears from her face, turned and marched to the door. She pulled it open and stepped through as her friends chanted, 'Good luck, Amie. Good luck.'

The school was virtually empty. Everyone had gone long ago. That was a sign, thought Amie. She didn't look left or right, just straight ahead. Don't look at anyone. Don't look for anything. Just walk and get home safely. She didn't see the security man at the school gate raise his hand as he always did. Didn't hear his 'Bye, Miss!' She didn't see the bunch of kids buying fish-balls at the shop opposite or hear them shout, 'Hi, Amie!' She was just blocking everything out. She didn't think 'Why aren't they hurrying home too?' because she knew they didn't know. Should she tell them? No, they'd not believe her. A waste of valuable time.

Had Amie paused and looked back, looked up at the school block, looked up at her class window which she'd been looking through a minute or so before, she would have seen two faces. Two faces she knew and loved, her best friends, both roaring with laughter and pointing at her, tears streaming down their faces.

On the MTR, Amie found a seat for the three stops she had to go and stared up at the train map opposite, willing the next dot – the next station – to light up.

'Excuse me.'

Her heart leaped. The man next to her was leaning towards her. His eyes were red – blood red – his chin unshaved, his teeth broken and stained. She stared in horror at him, frozen, unable to move. He pointed a dirty, wrinkled finger downwards. Was this the sign? Was this the global signal for an attack?

'You dropped your rail card.'

The voice was rough and broken. She stared at him, her mouth sagging open. She didn't move. She didn't speak.

'Girl, you dropped your rail card. Your Octopus.'

His finger – his gross ghoul's finger – was pointing downwards at the card lying on the floor by her left foot. In a spasm she lurched forward and snatched the card up, clutching it to her chest. Words wouldn't form in her mouth. She just stared at him.

The old man shook his head and slowly looked away, his eye-brows shrugging above his eyes. Amie knew he was just waiting for the right moment. Perhaps when the other people got off the train. But they were looking at her, frowning, disapproving. No, they were waiting too. The old man would attack and then they would join in! She had to get out!

As the train pulled into W-------, she leapt to her feet and hurtled through the sliding doors before they'd opened fully. She pushed against a man, knocking him into the door.

'Hey! Watch it!'

Her legs took her through the shuffling crowds, bumping, dodging, nearly tripping, her shoulder bag jarring others as she ran. Voices shouted at her. Maybe they were running after her. There was a queue at the ticket barrier but Amie couldn't wait. She had to keep moving. She dashed down the side of the queue and, with one hand on the barrier, vaulted over to the other side. A couple of voices shouted at her. A woman in an orange MTR uniform. Amie thought was at least two sizes too small, swung her yellow megaphone at her but the fugitive ducked and ran on towards the nearest exit.

She knew she was two stops short of home which meant at least another mile to go but at least she wasn't trapped in that underground box, that metal coffin, with those disgusting...ghouls.

The main street – King's Road – was busy. It was the road that ran through the heart of the Island carrying the life blood of the city. That thought made Amie a little sick. It seemed normal. Everyone was going about their business. Shopping. Cleaning. Burning paper outside the shops. That's how the ghouls worked obviously. Being normal. Giving the appearance of ordinary, everyday life. Not like in the movies, lumbering around with bits dropping off them, mumbling like their mouths were full of sticky rice. They looked normal to trap their victims. It made it all so easy. Anyone or everyone could be a ghoul.

Amie avoided the pavements and ran down the centre of the road, following the tram tracks, moving from tram stop island to tram stop island. A tram driver rang his bell furiously at her when she dodged across the bows of the large, looming tram. But Amie was deaf to the noise and the occasional shouts and car horns that echoed around her.

She was nearly into her home area – nearly safe! – when two policemen paced out of the crowds on the pavement and pointed at her.

'Stop! Stop right there!'

Any other time, if two police had told her to stop, Amie would have done so. Immediately. She'd been brought up to respect the law. But who knew? Perhaps the ghouls had infiltrated the police. These two could be ghoulstables. Taken to a cell, she would be at their mercy. The two officers stood, their spectacles glinting, hands out wide to block her way. Traffic was sliding past on either side, a shifting wall of steel.

'Stop! What the hell do you think you are doing?' said the taller man, spit flecking his chin.

Amie skidded to a halt in front of them, straddling the tram tracks. She stared at them. Her head throbbed. Her legs burned from the pain of running. Breath was coming in panting bursts.

'What's going on?' said the second officer moving towards her, his eyes unnaturally wide.

Amie tensed. There was no way back. Only forwards. Only home and safety.

Suddenly the air filled with clanging. Behind the officers a red tram loomed, the driver's face open mouthed. People on the pavement screamed. Amie yelled at the two policemen, 'Look out! Tram!'

The rattling, metal beast was being jolted by its screaming brakes. Amie saw passengers sliding about inside, hands clinging to the open window frames. The officers half turned and jumped sideways. Cars screeched to a halt. The sound of metal on metal shredded the air. The tinkle of broken glass. A mass 'Aaghhh!' from the pavements.

And Amie was gone. Sideways through the traffic, through the shouts and the car horns, through the barrage of yelling. Through the goggle-eyed people on the pavement, up a side-street, its steep slope barely slowing her down. She was running faster now, fuelled with adrenalin, charged on fear, brimming over with the need to get home, to be safe.

The noises behind her subsided. Now only her feet pounding the pavement, jarring her every part. Just to the end. Turn right and there were the apartments, Eternal Hope Block C. Never had she been so happy to see that sign.

Amie fumbled her key in the lock and burst into the flat. Mother and Father, sitting at the table, were sharing out a plate of noodles. They turned with puzzled faces as she staggered through the door, blinking in unison as she slammed it shut.

'You're late, Amie. Where have you been?' Her father's voice was chilly.

'I was…it was the…on the MTR there were…'

Explanations wouldn't come. That was when Amie burst into tears and collapsed into her mother's arms.

'Oh dear, Amie. What a mess up, eh?'

Amie's sobs shook the both of them. Mother's hand patted her back to calm her. The tears ebbed.

'Better now? Tell us all about it after dinner.'

Amie snuffled through tears and snot. It all seemed so stupid now. Perhaps none of this ghoul stuff was true. It was Dani and Zi having a laugh at her expense. She always believed, always trusted them. She was safe. Everything was how it always was. Not fake but real.

Father stepped round the table towards her. His arms went round her shoulders, holding her firmly, safely, upright.

'Come and have some dinner, Amie.'

His arm tightened on her. She looked into his face and he was smiling at her, his face shining with anticipation.

'What have we got for dinner tonight?' Amie asked, wiping her sleeve across her face to clean it.

Her father's grip on her shoulder tightened. His other hand had pinned her hand to her side. Mother reached down to the table and picked up the broad-bladed knife they always used for carving meat. She smiled into Amie's face but the smile was hideous: twisted, swollen, her eyes bulging with red veins, drool slithering across her chin. The knife was glinting in the light, held in a clenched fist above them.

'What's for dinner? Why you, of course.'

Dani and Zi had missed Amie for a couple of days. There was no news as to why she was absent from school. She hadn't phoned. Hadn't texted. Maybe it was the flu. Maybe she found a boyfriend, they joked. Maybe she can't do her Maths homework and is afraid to come in. How they laughed at the crazy reasons they made up for Amie's absence.

'Maybe the ghouls got her!' Chortles exploded from them.

It was the end of the last lesson. The bell had just gone. Zi's mobile buzzed in her pocket. She fished it out, swiped the screen and read the message.

'Zi, Dani. Come round now! Big srprz for U. HUGZ$FRNZ4EV.'

Doppelganger – the Double

You know how in the middle of a very ordinary day – a day like every other day when nothing happens – you suddenly become aware that something really weird – I mean REALLY WEIRD – is about to happen. It wasn't Friday the Thirteenth or Halloween or anything like that. No. A Tuesday. Yes, just a Tuesday.

There I was walking to school and yes, I was late again. The late teacher would be waiting with the detention book. I knew what was in store for me. I didn't care. One detention? Five detentions? So what? Just sat in a detention class with a teacher who didn't want to be there and me having the chance to annoy her by shuffling around, fidgeting, asking stupid questions. I mean, what is the point of detentions?

Anyway, before I go on any more, back to that walk to school. The street was empty – it was a back street – more like an alley really – that I always used. Empty except for one person at the other end heading my way. Head down, hands in pockets, bag on the back. Walking slowly. As he got nearer I could see he was in school uniform – my school – but heading the wrong way. Maybe, he'd been sent home. He'd got lucky then. A dude, maybe. Wearing the same trousers as me. And the same Nikes. A real dude.

His head was still down and we were just a few metres

apart and I realised he was even walking like me, kind of with toes pointing in a bit. Was he taking the mickey out of me?

'Hey! Aren't you going the wrong way? School is this way.'

My voice was laughing. He looked up. I started choking. My throat clammed up. My eyes were out on stalks like some goofy cartoon character.

The guy looked like me. Not just, like me. Was me. A perfect reflection. Everything exactly the same. The hair flopping on the forehead. The flat nose. Even the couple of spots on my chin (I need to cut down on fries – I know it). This guy WAS me. He'd stopped and was looking at me with a puzzled expression.

'What's the matter? Haven't you seen a student before?' he asked and even his sing-song voice was mine, rising slowly at the end of each sentence. I liked to think it always sounded sarcastic. Now I heard it and it sounded a bit phoney. I munched the air, trying to find words that made sense.

'Sorry…I…er…I…err…you…' and so on. Just jaw-dropped nonsense.

'And your point is?' His tone was aggressive. Like he wanted to beat me with his words.

'Well, you know, like, you look just like me. You look exactly like me. You could be…like…my twin.'

Me 2 looked down at himself.

'Yes, I guess I do. That is the idea.'

My mind fogged like a car windscreen.

'Eh? What idea? Whose idea? Your idea?'

He reached out and grabbed my shoulders.

'The idea. The plan. I'm your double. Your doppelganger. Your perfect copy. You're one of the lucky ones – cos you get to meet you. That's the game.'

'Game?' The mind-fug was swirling.

'Oh, it doesn't matter. I'll explain later. Okay, so here's the deal. You've been given one day with me and then I disappear forever. So what do you want to do?'

I reached and touched his arm. Yes, he was real. I held his shoulders so there we were, in the alley, a perfect match. Co-ordinated.

'Well? What do you want to do?'

My mind was suddenly full of ridiculous ideas. Things you only dream about. The chaos to be caused.

'Okay. Let's go to school. I'm...we're...late so there'll be trouble with the Late Teacher.'

Me2 laughed.

'Could be trouble for the Late Teacher.'

Late Teacher – LT with a badge to prove it – was in the entrance hall, scanning the list on his clip-board. With Me2, we'd already worked out our routine.

LT – Mr. Chin, the Maths teacher – looked at me and scowled.

'Lee. Five minutes late again. Why?'

'My dog got out and I had to find her.'

'Yesterday it was your cat got out. And the dog got out three times last week.' He checked the list. 'Monday. Wednesday. Friday.'

'My dog doesn't have a GPS fitted. He gets lost.'

Chin muttered something about wishing I did too. I didn't quite catch it. He mumbles a lot.

'Don't be funny, Lee. Keep the dog tied up. Maybe with the cat since they both like running around.'

'He doesn't like being tied up, Mr Chin. He bites the furniture.'

'So would I if I had to live with you. Detention. Get to class.'

'I must go to the toilet, sir. Excuse me.'

The toilet door was next to him, neatly labelled 'Boys' in the school colour. Blue. I went in and closed the door.

Cue Me2 through the entrance door. Mr Chin does a double take (I'm peeping through the crack in the door).

Me2: Sorry, I'm late, sir.

Chin: What are you up to? How did you…?

Me2: My dog got out and I had to…

Chin: I know about the blasted dog. How did you get out there?

Me2: I just walked in off the street, sir. Better get along to classes, hadn't I?

Chin: Hold on! Hold on, Lee. Not so quick. Are you pulling a fast one here? How did you…?

Me2: Do I get a detention slip?

Chin: You already have one, you idiot. You can't have two.

Me2: No, sir. I just got here now.

Chin: You're in the toilet! I saw you go in!

Me2: I've been coming here – to school. I haven't been to the toilet. You would have seen me.

Chin: I did! I did see you, just now. What is going on here? I've already put your name on the list.

Me2: Cross that one off. Excuse me, I have to go to the toilet.

And Me2 pushed past Chin who was staring at his list, at the entrance door, at the now closed toilet door. Chin took a step towards the door and I stepped out, walking quickly towards my class through a set of double doors.

'Thank you, Mr Chin.'

Chin was scribbling something out on his clip board when Me2 walked out.

'Thank you, Mr Chin.'

And he followed me through the doors. Chin had dropped his clipboard and his pen and he was peering at his watch. Chin was in Groundhog Day and heading towards meltdown.

Before we got to the form room, Me2 had struck a deal. He would go to lessons in my place. I would go and chill out somewhere close and we'd meet up at lunchtime behind the library block – no-one went there. So I spent the morning hidden away in a little store area, using my i-phone, sleeping, laughing to myself. This was fun. But what did Me2 mean when he had said 'the plan' – like someone had decided, someone had sent him? What was that all about?

Me2 wasn't there, of course, at the time we'd agreed. Nothing. No sign. I knew we couldn't be seen together – that would blow everything. But he wasn't around so I ventured down to the dining room. As soon as I came in through the doors, everyone's eyes turned on me.

Ali – one of my so-called friends – ran up to me.

'Lee. Where have you been? Why did you do it?'

My head was fogging up again.

'Do what? What have I done?'

'What have you done? What have you done? You attacked Mr Chai, the history teacher. You're in big trouble. Why did you do it?'

My scalp was moving round the surface of my head.

'I didn't do it. I didn't do anything.'

'Chai was really hurt. You hit him with a chair.'

'I did what?'

Hitting Chai with a chair was not me. Running him over with a tank – yes. But chair – no.

'You crept up behind him and smashed the chair over him. We all saw it. You can't deny it. But why?'

'But that wasn't me, Ali. That was…'

My tongue stopped working. Who would believe what I had to say?

'What did I do after?'

'You ran out. Oh, come on. How can you not remember? You said you had other scores to settle.'

'I said what?'

A dark cloud was beginning to descend as I realised just how out of control this had become. What had I told Me2 that he might act on? I racked my brains. Oh any number of things. Any number of people I didn't particularly like. But I wouldn't smash them with a chair.

Mizzy, a Year 3 girl, came running up with a gang of goggling friends.

'You're crazy, Lee, still being here. You should get out quick. They're calling the police.'

'Who is? Why? What are you talking about?'

'The Chemistry class. You know. You threw the acid bottle at Miss Chang.'

'I did what? But I didn't. I wasn't there…'

'It was you alright, Lee. You'd better get out quick.'

At the far end of the Dining Room, the double doors had been banged open. The Principal, her eyes on fire, was staring at me and pointing. Beside her were two police officers. As I turned to run, they began to give chase, shouting at me, 'Stop, boy! Lee! Stop right there!'

I wasn't waiting around to try and explain the impossible. Mizzy and Ali stood in front of them as they began running but jumped out of the way as the officers got to them – the cops were moving too quickly to try to slow them down.

I was beating tracks out of there at top speed. I knew where to go – the little back corridors that would take me behind the sports block, past the basketball courts, full speed. I wanted to get away but I wanted to get to Me2. What the hell did he think he was doing? This wasn't part of the deal – that he'd actually hurt people…badly. This was supposed to be fun. Not carnage.

I was off site. I'd scrambled over a wire fence, nearly broken my ankle falling down the other side and was now in a deserted play area between four tall tower blocks. It felt abandoned, empty to the point of desolation. I slumped onto a swing.

What the hell was I going to do? No-one would believe me if I said it was my double who I'd met on the way to school. No-one.

'No. They wouldn't believe you.'

My voice came from next to me. I turned my head and Me2 was sat on the next swing. A wave of sickness ran through me, finally sitting itself in my throat.

'What have you been doing?' My voice was strangely croaky and distant.

Me2 smiled at me but his eyes were gleaming dark.

'Just a bit of fun. That's what you said, wasn't it?'

My hand reached out and grabbed his throat.

'Fun! Not attacking people. Not hurting people like that.'

Me2 grabbed my hand and pulled it free – too easily. He seemed to be strong – very strong. Too strong. Not like me at all.

'You didn't like those two teachers. I saw your thoughts…'

My jaw dropped. Now he could read my mind. What else could he do?

'Don't look at me like that.' He was grinning – a thin grin that played along his lips like seaweed in water. 'Of course I can read your mind. I'm you, stupid!'

Everything wanted to explode inside me.

'That was my thoughts. My imagination! I couldn't actually do any of those things.'

'Well, I can do those things and I have. You wishes are my commands. And it's too late to do anything about it. You've let the real you out on the loose.'

In a second he was out of the swing seat and away across the park, leaping over the play equipment, away, away, away from me. Where was he going? What evil thing was he going to do?

Then it struck me. What evil thing had I thought? Whatever it was, he would act on it.

My so-called friends at school? No. He'd had the chance. No. But what about home? Had I had thoughts about…

My little brother, Aden. Eight years old and always annoying me. Coming into my room. Taking my things to play with. I'd told him to get lost. I'd wished he wasn't there. Yes, I had. Only last night.

Home was fifteen minutes away. I looked at my watch. It was Aden's morning school day. He went home in the afternoon. And Mother didn't get back till three to look after him. It would be Jin, the girl from the next flat, who'd be there with him.

I launched myself from the park. All thoughts of pursuing police had gone from my flaring brain cells. I had to get to Aden before that monster Me2. My feet pounded on the concrete pavements. My heart pounded in my chest. My brain pounded in my head. Please don't let any more bad things happen. Please. Perhaps the police would go there to catch me. That would be good, wouldn't it? They'd stop Me2 wouldn't they? Wouldn't they? And then when I arrived…all would be clear. Wouldn't it?

I was now into a busy street, dodging between people, stepping out into the road making taxi cabs hoot angrily. People shouted as I bashed into their shopping bags. But nothing was going to stop me. I had to get home before Me2. A red traffic light and a crowd had formed on the edge of the road. The red man shape on the light said: 'No'. I leaped forward, around the edge of the crowd. An engine roared and the red bus seemed to tower above me. Tyres screeched and people screamed. The bus driver's face seemed to open wide in slow motion as his white fists wrestled with the steering wheel. I threw myself forward and rolled onto the pavement opposite. A couple of men shouted angry words and reached out to grab me with their big, brown hands but I slipped past them like an eel.

Into our street at last. And there was our block – painted in peeling yellow. Decorated with hundreds of lines of washing. Home! How I hated it usually. But how I wanted it, needed it now. Seventh floor. I rushed to the lift.

'Out of order'

A hand written notice clinging by a couple of crumpled pieces of Sellotape. I banged on the metal doors and cursed. The noise seemed to echo up the building.

'That can't make it work, boy,' said a man kneeling by the door, fiddling with some wires on the wall. I jumped towards the stairs and pushed through. Seven floors. At least fourteen flights of steps.

The first four were easy. The next four were harder. I was slowing down. Now I started to make mistakes. Feet catching the step, falling forwards, pain arrowing up my shin. The last ones to go. My thigh muscles were burning and bunching up on me. My breathing was on fire in my lungs. Everything in my body was telling me to stop. But I couldn't. I had to get to Aden before Me2.

Floor Seven. I fumbled for my key, dropped it, tried to put it in the lock but my hand was shaking too much. I steadied the one hand with the other and slowly, finally slid the key home. One twist and the door swung open. I half fell to my knees as it swished away from me.

'Aden! Aden! Where are you?'

That's when I saw the blood on the floor. And the smashed window with a hole big enough for a body to fall through. I couldn't step forward to look through the hole. Couldn't bear to see Aden's body sprawled on the pavement below, smashed and distorted.

'Aden!' my voice howled. But the voice wasn't connected to my brain somehow. All this was happening outside of me, away from me, in another awful, horrible world.

'Yes, Lee. You're here.'

I turned and there was Aden in the bedroom doorway, smiling. I couldn't move. I was frozen to the carpet. Nothing made sense any more.

'You're safe!'

Then I rushed to him and threw my arms round him, something I had never done in his life. I wrapped him tight, pulling him to my chest, stifling his words.

'Of course…ah…arm…'

I held him at arm's length, looking into his eyes.

'Did someone come? Someone who looked like me?'

Aden smiled a wobbly smile.

'Yes, but he saved me.'

Brain fug again.

'Saved you? Who saved you? Where is Jin?'

Aden nodded towards the broken window.

'Who saved you, Aden? Who saved you?'

Aden smiled again and pointed his finger at a figure who now stood in his bedroom doorway. My flesh seemed to slide from my bones. Everything was now a red mist but through it I could see the figure, a halo of light around it. It was Aden.

World Book Day

'What a lovely idea,' thought Mrs Chang when the Headmaster announced that the whole school would celebrate World Book Day. She pushed her glasses back on her nose and clicked her pen excitedly.

'Perhaps, Headmaster,' she suggested, staring at his face to see what his reaction would be, 'the children could come to school dressed as characters from their favourite stories.'

The Headmaster smiled thinly. This was not the sort of thing he liked at all – joy or pleasure. 'A splendid idea, Mrs Chang. Perhaps you would like to co-ordinate it.'

Someone to her left in the staff meeting, sniggered.

'Well, yes, Headmaster. I'm more than happy to do that,' she replied, clicking her pen even more. 'And perhaps we teachers could come dressed up as well.'

Several people in the room moaned softly to her right. Sniggerer on the left sniggered again. The Headmaster shuffled his papers and coughed.

'I'll leave that for teachers to decide. Perhaps Mrs Chang, you could offer advice. Or even organise staff to dress up.'

More moans and sniggers. Mrs Chang looked quickly to her left but couldn't identify where the snigger was coming from.

'We don't want lots of teachers coming in dressed as Mickey Mouse or a Transformer,' the Head continued, his eyes shiny.

Mrs Chang shuffled on her seat. 'Oh, I wasn't thinking that Mickey Mouse or a Transformer would be considered figures from literature.'

'No, I'm sure they're not. Just examples. Unfortunately,' and was that a smile playing on his lips? 'I'm away on that day so won't be able to take part. But I would like to think I would come as Mr Pickwick from Charles Dickens' Pickwick Papers.'

Mrs Chang could think of no-one who would be less like Mister Pickwick than the Headmaster. Pickwick was funny, full of fun, intelligent, popular…but she clapped her hands and smiled. 'A good choice, Headmaster. Such a pity you won't be here. But it will be good fun, won't it, everyone?'

Mrs Chang looked round the circle of teachers' faces. There were not many signs of fun there. In fact most of them looked distinctly miserable. Even the sniggerer had stopped sniggering.

So that was it! World Book Day and dressing up it would be. What an exciting project, she thought.

The next day she posted a list of teachers' names on the staff notice-board for people to sign up to which characters they would be. She had already written in her choice: 'Jane Eyre'. The moody, strong-willed woman in a black dress who endured so much pain and hurt in her life. Yes, she was Jane Eyre.

At the end of the day only a few names had been added to the list.

Mr Bryan (English): Moana (well, he was from New Zealand)

Miss Wong (Chinese History): Dumbo (she did have big ears, thought Mrs Chang, so it would be okay)

Mr Ming (Deputy Head) : Jack the Ripper (although that didn't look like his handwriting)

Mrs Chang was puzzled and slightly angry. None of these were really literary figures. They were supposed to be from books. Not cartoon films or history. She decided she would have to explain that to staff the next day.

The next day came. She put up a fresh list and made it clear that the choices had to be from fiction. 'Made up stories' she added, just to be clear.

'But isn't history just made up stories, Mrs Chang?' challenged Mr Lee, the English teacher who always complained about food in the canteen. Mrs Chang shook her head.

'No, Mr Lee. History books are factual. We're talking about literature. Great books like…'

The Headmaster broke into what might become a long discussion about the nature of history.

'Yes, Mrs Chang is right. Keep it simple. Children's books are best. Ronald Dohl and all that.'

'Dahl. Roald Dahl,' corrected Mrs Chang.

'If you say so, Mrs Chang. Put your choices on the list.

Now, the duties for this week…'

Why were people looking at her so angrily? thought Mrs Chang. This was going to be fun.

When she told her class, they thought it would be fun. Apart from the Wang boy, Charles. But he was always, well, difficult. Like some of the staff.

'Could I be someone from African literature? Or Mexican literature?'

His eyes were boring into hers. Mrs Chang swallowed. She sensed another argument.

'Well, yes Charles, you could. But then we wouldn't know them, would we?'

'You would if you were African or Mexican.'

'Yes, but Charles, look around here. Do you see many Africans or Mexicans here? We all seem to be…Chinese.'

Charles interrupted again as she drew breath to continue.

'So why are we choosing characters from English Literature?'

Mrs Chang juggled a few ideas, a few arguments in her head then spoke clearly and firmly.

'Because we are.'

And that was it! World Book Day would be fun and a great success. What a pity she would have no-one to share that success. Mr Chang – her husband, well ex-husband – had left her and moved to Shanghai five years ago. A cruel man – not unlike Mr Rochester in 'Jane Eyre'. Now she could turn her attention to her costume.

Come the day and everything was perfect. The teachers rolled up, glum-faced as different characters: Bill Sykes (Oliver Twist), Anne of Green Gables, Mr Ming was Mister Twit ('Most appropriate' commented someone in the corner). But Mrs Chang's outfit was the most impressive. She'd worked hours, cutting, snipping, sewing. The black dress fitted perfectly with a row of pearl buttons down the front. The dress had several petticoats and flounces and spread out neatly round her ankles. It rustled as she walked ('like a bag of chips' she heard someone say jealously). The outfit was finished off with a black bonnet held with a black ribbon tied neatly under her chin. She was Jane Eyre.

'The Woman in Black?' someone asked.

'No, Jane Eyre.'

'Florence Nightingale?'

'No. Jane Eyre. Florence Nightingale is from history, Mr Pan.'

'Aunt Piker from James and the Giant Peach?' asked another.

'No. Jane Eyre. Proper literature,' replied Mrs Chang sniffily.

At least the staff had done it and come as literary figures, on the whole. Although Miss An looked a bit odd in her pink rabbit outfit, (Alice in Wonderland she had said). Someone said they'd seen her wearing it in a nightclub in Wanchai.

But the class! What valiant efforts they had made. Dorothy

from The Wizard of Oz. Harry Potter (five of them). Even the Child Catcher. Some Mister Men. Hermione (Harry Potter again). But a delightful selection.

When Mrs Chang called the register, Charles was missing. A little part of her was relieved.

'He's the Invisible Man', someone suggested unhelpfully.

But just as the bell was about to ring for the start of lessons, Charles appeared at the door. Mrs Chang gasped for breath. Charles was wrapped from head to foot in bandages. There were holes only for his eyes and mouth. Wires dangled from his head, his chest and his arms. There were what appeared to be bloodstains on his sleeves. The children fell silent as he seemed to stagger to the front.

'Charles!' Mrs Chang recovered and spoke too loudly, almost shouted. 'What an outfit! Who exactly are you?'

'Fraaaaonkonnstaffmarsty.'

Mrs Chang shook her head.

'Sorry, what did you say?'

Charles poked a finger through the hole where his mouth was.

'Frankenstein's Monster.'

Mrs Chang stepped back.

'Oh, I see. Very good. The Mary Shelley book, Frankenstein. Why didn't you come as Professor Frankenstein, Charles?'

'...ooorin...'

'What?'

Finger in mouth hole.

'Boring.'

'Oh, I see. Well, it's jolly…different. But won't you find it difficult to write and move about?'

'Thasherodee.'

'Pardon?'

Finger poke.

'That's the idea.'

'Oh, I see.' But she didn't.

'Miss Chang, wasn't Frankenstein's monster horrible? Didn't he…' enquired Little Miss Naughty.

'Well,' began the teacher.

'Yes,' interrupted Professor Brainstawm. 'He murdered loads of people. Including children. He used to pull their…'

'That's enough. Quite enough. The monster was… perhaps… misunderstood.'

'I wanted to kill those children'

Charles' voice was suddenly clearer. Louder. Mrs Chang scanned the white, bandaged face but could see no glimmer of life in any of the three black holes.

'Now, children. English Grammar time. Take out your pens. Mary Poppins – can you help the Monster with his writing?'

Mary Poppins smiled widely and put an arm round the Monster.

'I'll look after him, Mrs Chang.'

The morning went quickly. At coffee break, Mrs Chang felt everyone was avoiding speaking to her. Perhaps her Jane Eyre outfit was a bit off-putting. She did look very serious and severe in it. Only Miss Tang ('Mary Magdalene', well, Miss Tang was very Christian so 'the Book' was justified) spoke to her about how she'd made her biblical outfit from tea towels. Mrs Chang didn't feel it would be useful to talk about the stories that suggested Mary Magdalene had been a prostitute but would have liked to spend a bit more time talking about the stitching (rather untidy in her opinion) but the bell went and everyone pushed out of the staff room.

After break, Mrs Chang was a bit worried to find that Cindy (Mary Poppins) had not returned to class.

'Has anyone seen Cindy, who, as you know, is dressed as Mary Poppins. She should be here.'

'Looking for an umbrella?' suggested one of the Harry Potters.

The Monster shuffled to his seat.

'I think she had a doctor's appointment. I saw her leaving.'

Mrs Chang sucked on her pen.

'She didn't bring a note as she should. Thank you, Mons… Charles'

Lunchtime break soon arrived. The children had been excited except for Charles (Monster) who seemed to have shut himself off inside his outfit. He hurried out as fast as his bandaged legs would carry him. Several children from younger classes screamed as he went past.

'His outfit is very realistic,' mused Mrs Chang, putting the English text books in neat piles in the book-case.

She ate alone in the classroom as she usually did. Not unlike Jane Eyre herself, she thought. Jane Eyre – educated, sharp-witted, ambitious yet caring and loving. The choice was obvious to Mrs Chang. She paused in the middle of her char-sui. Was there something about the taste that wasn't quite right? A little bitter? Like burned almonds? Or was it her imagination? She coughed a little and couldn't finish the portion.

The children trooped in for the afternoon session, a little reluctantly as they'd obviously had great fun at lunchtime. Some of the outfits had tears or pieces missing. Hats had been exchanged. They looked more like a bunch of pirates who'd pillaged a local town than a group of literary-minded school children.

Mrs Chang took the register.

'May Ling. Has anyone seen May Ling? She was here this morning. She was Robinson Crusoe.'

'Perhaps she's gone because today's Friday,' popped up Paddington Bear.

'Don't be silly. It's Wednesday,' snapped Mrs Chang. She suddenly wasn't in the mood for funny comments.

'Perhaps she's just gone.'

The children turned to where the voice had come from. The Monster. A silence fell. There had been something…odd about Charles' voice.

'What do you mean, Charles? Gone? Gone where? Gone when?'

The Monster slowly shrugged its shoulders. Mrs Chang coughed again. Now she felt a searing pain run through her stomach. It made her dizzy. She clamped her hand to her desk to prevent falling over.

'Are you alright, Miss Chang? You look ill,' chirped a cross between Huckleberry Finn and Gandalf.

Mrs Chang steadied herself and the dizziness passed. She snatched at the bottle of water on her desk and swigged at it. She had a raging thirst. A dribble of water ran down her chin and onto the white frill round her neck.

She knew she should check with the office about May Ling's absence. A quick phone call or she could just go up the corridor. But she couldn't quite remember what she had to do. She was feeling confused.

She stared at her planner. The afternoon lesson said something about…'something…Science' but she couldn't make out what it was she had intended to do. In front of her the children were getting restless, not sure of what to do, what to say, with her swaying at the front of the room.

'Do you want to sit down, Mrs Chang?'

'Shall I go and get someone, Mrs Chang?'

'Are you alright, Mrs Chang?'

Mrs Chang slapped her hands on the desk loudly. The children jumped and there was silence. All eyes were on her. Except the Monster, whose head was turned to the window, staring out it seemed, at the distant mountains where black clouds were massing, his head on one side, as if listening for something. There was a rumble of thunder.

'No. No, thank you. I'll be okay in a minute. Just bear with me.'

Her mind was muddied. Somehow, the lightness and joy – the fun – had got lost somewhere. Her vision cleared and she could now see to write on the board the instructions for the afternoon. It was about 'Electrolysis' or something like that. So fuzzy. So strange. She was always so sharp.

Somehow – she wasn't clear where or when – she dropped into a sleep. It may have only been for a few seconds but it was a sleep peopled with strange figures that shouted at her, pulled at her, demanded this and that, expecting her to do this…do that. A terrible confusion – all the things she hated.

Suddenly her eyes sprang open and she was awake. She was still feeling sick and there was a nagging pain in her stomach. But the class was quiet. Here they all were, sitting facing the front, silent.

'Oh, I'm sorry, children. I drifted off for a moment.'

'Forty minutes.'

The voice was harsh. Accusing. And it came from the hole in the face of Charles. It was not his voice. Not Charles' voice. But an adult voice. A much older voice. Was this some sort of prank? To humiliate her?

'Charles. What are you doing out the front?'

Charles was in the teacher's position. On the desk in front

of him appeared to be electrical equipment – dials, wires, liquids – all organised around a large, central box with a thick cable that ran to the window and then outside.

'What is all this…stuff, Charles?'

Mrs Chang's voice sounded wobbly. All the authority had gone.

The Monster raised a hand to bid her be quiet. He pointed to the window. She turned to see how dark it had become outside. It was like the darkness spreading within her. The black thunder clouds has flattened out all over the city, a dark blanket suffocating everything beneath. Across the city forks of white lightning flashed.

'It's going to be a storm, Charles. Better close the window. The rain will get in.'

The Monster shook his head.

'No, I can't do that, Mrs Chang. I need it open.'

That voice again. She looked around at the children. Still no-one spoke. They all stared at the Monster at the front.

'Children! What are you doing? Why aren't you working?'

No response.

'It's called mesmerism, Mrs Chang. A science experiment I ran while you were…out. They are all mesmerised. They can't do anything.'

Mrs Chang's mouth opened but no words came out, at first. Finally they blurted out, in a rush.

'You can't do this, Charles. Running an experiment…with the children. You're not allowed. I am the teacher. You can't do this.'

'I can and I have. And all we are waiting for is the lightning.'

The deep adult voice was tinged with childish humour, childish glee.

'I wanted to make our World Book Day more realistic. More interesting. More shocking.'

At the moment that she realised she was held, restrained by plastic ties, her hands and feet bound to her seat, Mrs Chang saw the wires. Each child had a wire clutched in their fist, the wires running forward to the gadget on the desk in front of Charles and that box was connected to the cable that ran to the window.

'Charles! What are you doing? Why have you tied me up? You can't do this. Stop right now! If the lightning strikes the cable…they'll all be electrocuted. They'll all…'

The Monster shook his head and indicated the table in the corner.

'Not all, Mrs Chang. They won't all die. Some will come back to life.'

She turned to look at the table. The bodies of Cindy and May Ling were stretched out on it. More wires were fixed to their heads and bodies. The Monster seemed to shake with excitement.

'You killed them! You killed Cindy and May Ling!' Anger now burned instead of the sickness inside Mrs Chang. 'What are you doing? This is madness.'

'They will wake up to find everyone else dead.'

Mrs Chang felt her whole body go cold – every inch of it. The ice penetrated her heart, her mind, her soul.

'How can you do this, Charles? Charles? Who are you? You're not Charles, are you? Who are you? What are you doing? What evil is this?'

She looked down at herself. There were no wires connected to her.

'You haven't connected me, Charles. Whoever you are. Why not? What are you trying to do? What does all this prove? This is evil.'

The Monster stood before her, defiant, bandaged hands resting on bandaged hips.

'Not evil, Mrs Chang. Imagination. Imagination. You remember that, don't you?'

A clap of thunder over head, shook the building. Lightning fizzed nearer now.

'But I don't understand!' The teacher's voice was now a wail.

She tugged helplessly at the plastic ties that held her in her seat.

'Mrs Chang, you wanted the world of books. You wanted stories and wild characters and fantasy! Stories of death and daring! Well, this is the Dark Imagination. This is Mary Shelley's mind. And Bram Stoker's. And Edgar Allan Poe's. And Stephen King's!'

Thunder rattled the windows in their frames. None of the children reacted. Not a flicker. The teacher strained against her bonds.

'You can't do this, Charles. You must not do this! You will

have it on your conscience forever. Killing all these innocents.'

'Just like in the stories. The poems. The Pied Piper. He took the children. The Child Stealer. He took the children. Now Frankenstein's monster will do what he was designed for. Be a monster!'

Mrs Chang's eyes bulged with disbelief. This couldn't be happening.

'But you're not the Monster. You're Charles. A schoolboy.'

'Ah, the power of the imagination, eh? Oh no, Mrs Chang, I am the Monster. Believe me. And you will be the only witness to tell the story. You will become the most famous, the most notorious story teller of all.'

Again the chair rocked with the teacher's efforts to be free. The Monster's voice was calm.

'You will live to tell the tale. But not for long. The poison will take you in a few hours. Enjoy your fame while you can.'

She looked down at her stomach and felt the burning sensation again.

'Why, Charles, why? Do you hate us all so much?'

'Imagination became reality, that's all. I am no longer Charles. I am the Monster. And this…'

He moved quickly to the window.

'…is the storm which will despatch us.'

'Us?'

He held up the wire in his own hand.

'Oh yes! Us! Like the Monster I couldn't live with what I've done. Nor could Frankenstein. So goodbye, dear world…'

The thunder now roared like a beast above them. There was the sound of falling bricks and slates, breaking windows, electric lines snapping and humming. A sudden huge gust of wind tugged at the open window like a wild hand, tearing it from its hinges.

Mrs Chang screamed as glass fragments bounced across the room, cutting her face.

Then came the lightning flash. Blue. White. Orange. An enormous crack of a whip. A buzzing that bored into the brain. The sharp, nauseating smell of hot metal, of burning, of burning, of burning flesh. And over all that bedlam, Mrs Chang's scream.

An alternative…

If you find that ending just too…disturbing, then read on for an easier ending.

'What a lovely idea,' thought Mrs Chang when the Headmaster announced the school would celebrate World Book Day. She pushed her glasses back on her nose and clicked her pen excitedly. The she remembered. The dream.

'Perhaps, Headmaster, that is not such a good idea. You see I had this terrible dream about World Book Day and everything was terribly wrong and…'

The Headmaster smiled thinly.

'Don't worry, Mrs Chang. I'll be organising it. Nothing can possibly go wrong. It just needs a bit of imagination.'

There was something about that voice that made Mrs Chang shiver.

Notes:

1) Mary Shelley's book, Frankenstein was written in 1818 by the eighteen year old Shelley. Doctor Frankenstein creates a living being which turns out to be a murderous monster.

2) Bram Stoker wrote Dracula, one of the original vampire stories, in 1897.

3) Edgar Allan Poe: a nineteenth century writer who specialised in horror short stories like 'The Fall of the House Usher', 'Murders in the Rue Morgue' and 'The Pit and the Pendulum'.

4) Stephen King: A modern American writer who has produced a string of popular horror novels including Carrie, Misery and Salem's Lot.

5) Other horror writers: H.P. Lovecraft, Sheridan Le Fanu, James Herbert.

The Curse

My name is of no note. In America, I would probably be called Bill – but not here. Well, actually, I am one hundred dollars. But that isn't a name, is it? It's just what I'm worth. Sometimes people who have me call me 'Easy'. I'm just 'Easy Money'. Others say I'm 'Hard Earned' – which must be the opposite. But to everyone who holds me, I'm 'Okay' in their eyes. But I'm not so sure. I think I have a hex on me – you know, some sort of curse which brings bad luck. Bad things happen to people who have me. All sorts of things like the man who…

No, I'm getting ahead of myself. Let's begin at the beginning. A very good place to start. A cash point somewhere in the city. My moment of birth – emerging into the world. I slide out through a slot with four of my friends. No doctors. No midwives. An easy birth.

My first owner is a man. A smiling man who counts us, one, two, three, four, five. Whoops! The wind is strong on this day. Very strong. It swirls and sweeps through the spaces between the sky scrapers, whipping up dust, papers and… me. This one hundred dollar bill is tugged from the man's pudgy fingers as he clumsily tries to push five notes into his wallet.

I am spinning away, landing on the pavement then

flipping, over and over in a crazy dance away from the man and his clutching hand. He is running after me now, sometimes trying to catch me with his hand, sometimes trying to stamp on me but to no avail. It is a pointless chase in the street. Maybe he should be shouting 'Stop thief!' at the wind. He is swearing. He is sweating. He is angry and humiliated. People stare at him as he lunges past. No-one tries to catch me for him. Perhaps they think he is just a madman rushing along the pavement. I veer right suddenly, carried by a cross-wind into the road. The man steps after me, not looking, not taking care as he should.

A car horn blares. Brakes screech. There is a sickening thump and I see the man somersault through the air, heels over head then crash bone-crunchily into the gutter. There are a couple of screams. A shout. Then his wail of pain. I catch sight of his leg at a funny angle and something like a broken stick poking out of his trouser leg. He is screaming now. People run hard towards him. Running away from him too, unable to look at his pain.

I lose sight of him among the cars and buses that have now stopped. I shuffle along the road between them and flip over and over into the park opposite where the wind leaves me stuck flat against a rubbish bin. Just time to recover my breath – if I had any, which I don't. No-one notices me there for a while. People in the park are too busy dropping litter everywhere to notice a one hundred dollar bill by the bin. Then I hear someone coughing, coming closer. A rasping cough that talks of too many cigarettes, too little care. A cough that rattles in the throat and bubbles in the rib cage.

A hand, dirty and creased, reaches down and grabs me.

'Hey! My lucky day!'

The man is roughly dressed: a creased and dirty t-shirt, ragged jeans. He doesn't wash much – the creases in his face are lined with grime. He seems in a bit of a bad way, as they say. He's not a man who uses a cash point (or a bath come to that), but he is grinning and holding me up to the light to see if I'm real or just a Monopoly note. I am genuine, I want to tell him but I'm a hundred dollar bill and I don't talk. Although some people say 'Money talks,' I don't. The old man is delighted. Happy, smiley face. Finding me has made his day. I feel wanted and loved – good feeling.

He carefully folds me and, looking round, tucks me in his back pocket. It's a bit smelly in there – sweat and the kind of smell which means these jeans have been on his body for a long time. If I had breath, I'd hold it for a while.

The man hums and sings to himself, tunelessly, as he walks through the park. Then suddenly he stops and turns away, looking down at his feet. No more humming.

'Hey, old man. What are you singing about?'

Two young guys in jeans stand in front of him. He doesn't look at them. He doesn't answer them. They step closer.

'We've been watching you. You found something, didn't you?'

One of them grabs his t-shirt but he pulls away.

'Don't know what you're talking about. You should show an old man respect. Respect! Now let me go. I've got nothing.'

Now they are on either side of him so he can't get away. He looks round, his eyes rolling, for someone in the park to come and help. Maybe someone will if he yells. He takes a big breath but the young man behind him sees what he is going to do and wraps his hand round the old man's mouth and nose. He is snuffling and gargling.

'What did you find? We saw you pick it up and put it in… your pocket.'

The youngster reaches into the man's back pocket and pulls out…me! Discovery! Oh dear. I'd really like to stay with the old man. He likes me. Appreciates me and I don't think these two will.

'Hey, Chi! It's a hundred dollar note! It's our lucky day!'

The youngster holds me under the old man's nose. The old man tries to snatch at me but I'm whisked out of his reach.

'Oh no, you don't. This is ours now.'

The old man snuffles some more from behind the hand on his mouth.

'What's that, old man? Did you just say 'Thank you'? I thought so.'

The old man's eyes are rolling and glaring, like he's having a major meltdown. He slides his head up a bit and bites down hard into the fingers on his mouth. The young man howls in pain and releases him.

The old man has his voice back and bellows, 'That's my hundred dollars! Give it back!'

The bleeding fingers form into a fist and land hard on the man's nose, twice, fast. Blood spurts out down across his lips and he coughs. The other young man standing in front of him, punches him once, twice in the stomach, bending him double. He falls to his knees groaning. The first man knees him hard in the head, the second stamps on his back, pushing him flat on the ground. I'm now in the young man's pocket. I can't see but they're still doing things to the old man. Nasty things. Violent things. Things that don't sound good. Things that sound of pain and suffering. If I had ears, I'd cover them up. This isn't good.

It's a bit later. The two young men are called Chi and Dan-e. I listen to their talk. What they are going to do that night? I hope they get rid of me quick. I don't like their talk. It seems very angry all the time. Talk of 'the Boss' and 'revenge' and 'slicing'. Then suddenly it gets angrier.

'We gonna share the money, right?'

'Depends, man. I saw the old man first. I saw him hide it. I knew what he'd got.'

'No, you didn't. Not till we took the note out. We share it. We both found it.'

The first young man is holding me out now. I can see where we are. A little, dirty flat which has broken furniture and piles of rubbish and bottles. Not a nice place.

'This is mine. I'll buy some stuff we can share, maybe.'

'No, no friend. We share it equal.'

The next moment all hell breaks loose. Chi and Dan-E are fighting but not fists. It is knives. They are jabbing and thrusting and shouting. The noise is horrible – so full of anger and hatred. Chi drives at Dan-E who screams and falls. There is blood everywhere. Dan-E sits down, holding his stomach and staring, looking puzzled as if to say, 'What's happened? Where is this blood coming from?' Chi has fallen back into one of the battered chairs. He is having trouble breathing. He is sobbing now.

'Sorry, Dan-E. Sorry.'

Dan-E doesn't say anything but looks up at Chi then back at his stomach which is oozing blood.

Chi's hand is holding me tight. His fingers are sticky. There is a bad smell in the room. Bad stuff is happening and somehow, I'm the cause.

'Get them to an ambulance. Quick!'

A policeman in uniform is bending over each of the young men and talking to an ambulance officer in the doorway. She has quickly bandaged the two fighters and with another officer, they carry Chi out on a stretcher.

'This one's gone,' says the policeman, shaking his head. He opens Chi's hand and picks me out, wiping me on Chi's t-shirt.

'Huh! Maybe they were fighting over this.'

If I could talk, I would be a good witness. I could tell them everything. But I can't, so the world whirls on, making up its own stories about what happened.

This is the first time I've been in a police station. I've been spread out on a table with some keys, a knife and a wallet. I think they are Chi's. A man is making a list of what is around me, including me. He talks to himself – he is alone.

'One knife. Switchblade. Nasty. One wallet with two ID cards. Some keys – three. One is the door key. And a hundred dollar note.'

He picks me up. I want to tell him not to do it. I'm no good for you! Things will go wrong.

Leave me, I want to tell him. Put me back. Throw me away. Anything. But it is too late. He folds me and puts me in his back pocket.

It is a gambling house when I appear again. The man from the police station is sitting with six other men round a table. They all have playing cards in their hands and stacks of money in front of them. Except my man. He is holding me up between two fingers. His face is sweaty and he is looking round at the other men. He is smiling at them but they are not smiling back.

'Okay! My last hundred bucks, guys. Okay? My last hundred.'

'If you lose, Wang, you owe us all a thousand. You get it, eh?'

Wang – my man – nods. I want to tell him: 'Don't do it, Wang. I am not lucky. You should see the other guys who got me. I will bring you…'

But before I can finish my thought, my warning, he presses me onto the table.

'One hundred sees you. A pair of jacks.'

I don't know what it means but it's obvious he's lost. The others throw down their cards and one of them scoops up the money in the middle, including me. The others stare up at the ceiling and then at Wang who slumps in his seat, his head down. Like when Dan-E and Chi died. But he's not dead. Not yet. One lip is quivering and he doesn't look well. A greasy pallor fills his face.

'Five thousand, Wang. Five Thousand.'

Wang doesn't look up.

'I haven't got it but I'll get it.'

His voice is weak and thin, like a child's.

'Of course you will. You work for the cops. Ask them for a loan.'

All the others laugh. Wang doesn't.

'We'll give you till tomorrow morning.'

Now Wang stares bug-eyed at them.

'You know I can't get that much…not so quickly as that. I'll need several days. Maybe a week.'

'Shut up, Wang. Tomorrow morning. Better get busy.'

The voice clearly means business. It is as cold as an ice cube on a hot neck.

Wang staggers to his feet, looks at them all once more and struggles towards the door. When he's gone, the others burst out laughing and bang the table with their fists.

'And all that with a lousy one hundred dollar bill.'

The man who owns me, holds me up and flips me round.

'Who's the lucky guy, then?'

I'd like to tell him he's made a big mistake. That he needs me like a hen needs teeth. But I can't. Because I don't talk.

There is a saying that 'Money is the root of all evil.' But I'm not evil. I'm just neutral. A thing. An object. It's what people do with me that counts. I don't want them to do crazy things. I don't force them to be bad. They just act that way. Maybe, if I wasn't here, things would be okay. But a world without money? I can't imagine it.

Gambler guy is called Ping. I hear the others round the table call him that. Their sharp eyes and tight mouths and tongues flicking across their mouths like lizards, tell me they don't like him. Another man appears at the table. The others shrink away. He is big – he blots out a lot of light. He wears a shiny brown suit. His teeth are like tombstones – yellow and broken. His voice comes from somewhere inside his monster chest.

'Hey, Ping. I'll play you for the hundred dollars.'

Why do people want to struggle over owning me? I'm just a bit of paper with printing on. Come on, guys! Ping scans the faces all round and licks his lips nervously.

'No, it's okay, Li. I'm on my way home. I'm out of here.'

Ping rises from his seat but Li's hand, like a fat leg of pork, pushes him back down.

'You weren't in such a hurry a minute ago. Sit down. Let's play.'

Ping's eyes are darting round now and sweat appears on his top lip in tiny bubbles. Then on his brow. Like his skin is turning inside out. His whole face is shiny with panic.

'W-w-what are we going to p-p-p-play, Li? P — p-poker?'

Big Li lets out a roar of laughter that makes everyone duck their heads into their shoulders like tortoises.

'No! Not poker! A proper gamble. Russian roulette.'

He laughs again and produces a revolver from an inside pocket in his jacket and slams it on the table.

'You know how to play Russian roulette, don't you?'

Ping is falling apart rapidly. The grinning winner of a few minutes before looks like a complete loser now, deflated and shrivelled, wishing he could shrink to the size of a cockroach and crawl out of the room.

'Y-Yes, Li. But…it's kind of dangerous.'

Li laughs again. It is not the nicest habit the way he laughs. It grates on people's nerves, especially Ping's.

'Of course it's dangerous. That's why it's fun. But it's only dangerous if you're not careful. Remember, when you spin the gun chamber, the weight of the bullet always takes it to the bottom. Well, nearly always.'

The fat hands work swiftly, opening the revolver, dropping the bullets out of the chambers, one by one, till only one remains. The revolver is snapped shut.

'You know the rules, Ping. Just one spin of the chamber. Barrel to the forehead. And pull the trigger. What could be simpler?'

Ping shakes his head. His hands are trembling.

'Li. I don't want to do this. I really don't want to do this.'

Fires burn hungrily in Li's eyes. He snatches up the revolver and presses the barrel against Ping's nose, pressing it back and up like a pig's snout.

'You have no choice, Ping. You're playing with the big boys now. Now lay your bet.'

Li snatches me from Ping's trembling fingers and slams me on the table.

'Here's my hundred,' he says. Grinning.

A note is banged down on top of me. Li leans his huge face towards Ping.

'My game. I'll go first.'

Pings nods and slumps, not wanting to watch but unable to look away. Li lifts the revolver, spins the chamber. He pauses, relishing the moment with everyone looking at him in horror. He presses the barrel to the side of his head and without pausing pulls the trigger.

Click!

'Boom!' he shouts and everyone jumps. He laughs. 'Now your turn, Ping.'

Ping reaches out for the pistol, his hand quivering in the air above it. Seconds become minutes, minutes hours, hours…

'Come on, Ping. Play the game,' rasps Li, leaning further forward as if to get a better view.

Ping grabs the revolver, spins the chamber and puts the barrel to his forehead. His finger goes white as it squeezes the trigger.

Now, you see for sure, I'm not exaggerating when I say I bring bad luck to everyone who owns me. At the moment that the gun went off, I became Li's property strictly. Now the thing is, Ping's hand was trembling a lot and at the crucial moment the barrel slipped off his slippery forehead. The bullet in its hurry to get out and create mayhem, slid across Ping's forehead, gouging a red line that sprayed blood. From there the bullet hurried across the room, glancing off a microwave in the corner, whining across the room to a metal safe in the wall then bouncing back towards the table. Li didn't see it coming, it was all so quick. There was a dull thud as the missile struck the back of Li's head and Li juddered forward. He straightened up with a very surprised look on his face. His jaw dropped as if he was going to offer more advice on how to play. Blood suddenly gushed from his nose and with eyes staring, he pitched forward, his head crashing onto the table. Everyone looked away not wanting to see the black hole in the back of his skull, fountaining blood.

Ping gasped, running his sleeve across the gash on his forehead. The rest were already running for the door, no longer wanting to be part of proceedings. Ping was finding it hard to breathe as he slowly stood up, pushing his chair over with a crash. Somehow he had survived. The gun was still stuck to his hand. He let it drop and looked at it in disbelief as it lay next to the two hundred dollar bills. My companion and I.

There is no underestimating the stupidity or greed of some people. Despite everything that had just happened, Ping reached down and grabbed up the two bills. The smoke from the gun was hanging in the air. There was a fine spray of blood settling around the room. Ping scrambled towards the door, his legs all jelly.

So here I am in Ping's flat. He has a little balcony overlooking the city. In the harbour way down there, lights twinkle. A light wind blows. The urgent whine of police cars and ambulances tingle the air. Ping is drinking a large glass of whiskey. He laughs to himself as he stands in the balcony doorway, holding me up in one hand. His other hand reaches down to the bowl of peanuts he's put on the balcony table. He is still laughing as he posts the peanuts from his fat, flat hand into his mouth. Now the laughing is coughing. Coughing and spluttering. He reaches for his drink, thinking he can wash the peanuts down but his hand, in its rush, misses and slides the glass off the table. It smashes into pieces. He tries to curse then tries to cough the peanuts free but nothing happened. He just dribbles helplessly.

He turns to run – perhaps to the kitchen- to get water but he is already giddy. The whiskey has had its effect. He trips against the table leg and falls on top of it. His face is red and swollen and his fingers tear at his throat, trying to get at the peanuts from the outside. His eyes are wide, blood-shot, popping from his fat face.

You see? I'm bad news for everyone. And as he's kicking and twitching on the balcony floor, that light breeze has plucked me from his fingers. I'm floating and fluttering out across the city. Hey, down there! Don't anyone pick me up. I'm your worst nightmare. I am the End for you. The Big Exit! DO NOT PICK ME UP!

The street is swinging closer. I can hear party music. There is a bar and light and noise and now I'm on a damp pavement, stuck to the drain grille. It feels like I will fall through. The best thing that can happen then no-one will be hurt anymore. I'll just slide into the darkness of this drain forever. A suitable ending. Down the drain, forever.

'Hey, look! A hundred dollar note!'

A young woman is stepping towards me. She is laughing with her friends. She thinks she is lucky. Tell her someone! Tell her to stop or this will go on and…

'I'm the lucky one!'

The Pickers

The following story is based on true events that happened in the UK in recent times.

Morecombe Bay. England. Twenty First Century.

Hu Wei knew something was wrong. None of the others had noticed but he'd lived by the sea. He knew what the sea could do and this wasn't good. This was very bad indeed.

He looked around the bay. The sands they were on stretched for miles to the rise of the cliffs in one direction and miles to the thin line of grey water that was the North Sea. That was where the wind was suddenly coming from – the sea. It battered at their faces and tugged at their clothes. He stood upright – the only one not bent, scratching the shell fish from the sand. Two things had changed – the sudden wind and the sand turning darker out towards the sea.

'What's the matter, Hu? It's not tea break yet!' laughed Li, his best friend among these twenty pickers. 'You Cockle Pickers' as the English boss called them in his strange, hard voice.

But Wei didn't laugh. His eyes were fixed on the area of darkening sand.

'There's something wrong, Li.' He answered. Li threw a big lump of wet sand at his bare legs and it splattered all over his knees.

'Get on with it, man. It's another hour before the boss man comes to pick us up. Get that bag filled or you'll be in trouble.'

Wei knew all about trouble. It had followed him since he was a little boy. At school the teachers had beaten him. Not for any real reason. They liked beating him because he was big and strong and they could knock him down. Then, after school, when he began working in the factory, the foreman had made trouble for him. Blaming him when things went wrong. Over and over – accusing him. Not working properly. Stealing. Being late. Being early. It didn't matter. Anything to make Wei's life bad. In the end Wei had broken and had lashed out with his fist at the ugly face of the foreman, bashing it, mashing it into a red pulp.

No more work at the factory. So he had gone back to his village, back to his mother and father, back to the fishing. But there was no money. The fish were being poisoned by the stuff in the water that flowed from the factory. The poisons and chemicals that made the fish float upside down on the surface of the sea. There was nothing for him and the family – they lived on what they could beg or grow in their meagre bit of garden.

That was when the man came and offered Wei a chance – a way out of all this. A way to help his family. The man said there was work in a place called England – a long way away. Thousands of miles. But it was okay. He could talk to his family every day. The man showed them two shiny mobile phones. All free! He would be able to send money home – thousands every month. Everything would be fine. In five years he could come home again – a rich man. Buy a house. Buy a boat. Buy anything. Even a good wife. The man smiled a lot and smacked him on the back and Wei signed his piece of paper.

Before he could even think about it, Wei was in the back of a lorry with his bag, his phone and a heart full of hopes. His school friend, Li, was with him. That made him feel good. They laughed a lot together – at themselves and at the growing band of young men joining them in the lorry. How they laughed!

'Hey! Wei! Come on, man. Get picking. Boss man could be watching us with binoculars!' Li's voice wasn't laughing now. Wei looked down at the sand beneath his feet. The ridges which had dug into his feet as he shuffled along were now softer. The sand felt spongy.

'The tide's changed, Li. The tide's changed. It's coming in.'

Li, heaved his bag, heavy with cockles, further up his back.

'Of course it will come in. When the waters get here, we'll move. Now get on with it, man!'

But still Wei's eyes were fixed on the sands. The whole shape of the beach seemed to be changing. Where there had been banks and hollows of sand, it was suddenly flat. The tide was coming in, but not as a wave. It was seeping up through the sand. Along each side of the bar of sand, they worked on; the sands were glistening with wet and even as he watched, little rivulets of water appeared, quickly becoming streams that rushed over the sands towards the beach, a mile away.

Wei's voice cracked with fear. 'We've got to get out of here! We've got to get back!'

More of the men stopped and looked at him. There was something in his voice that made them take notice.

'Look! The tide is already big!'

He pointed to the waters that now flowed around their sand bank, as fast as a river, waves folding over the top of other waves, a rushing torrent. Everything was changing as they watched.

Now they all stood and watched and fear was in their eyes. There was a wide river gushing sea water all around them. Their safe, solid sand bank was now a shrinking island. One of the men rushed forward and plunged into the water.

'I can't swim,' he yelled, 'I'm getting back now!'

But the tumbling waters plucked at his legs and he fell, the heavy bag on his back pulling him down. Wei jumped forward into the water and grabbed at the man's flailing arms but the sand under his feet sucked at his legs and he, too, fell forward. Only his strong arms and shoulders saved him as he beat against the water and pulled himself back on to the sand bar.

The others were shouting and waving their arms at their friend still in the water. He was shouting back but in a moment was pulled under as if a hand had grabbed him. He was gone. Wei got to his feet and all the men now dropped their bags of shell fish at their feet. They all seemed to be looking at Wei for help. Their eyes were wide. Their mouths open but silent, gulping in the salty air, gulping in what was happening to them. Wei looked towards the beach but the light was fading fast. Maybe the boss man was there. He would see and get a boat for them. A few lights were going on in the village farther along the coast. Maybe someone there would know what was happening…

Wei reached into his pocket. He always carried that mobile phone. Always phoned Ma and Pa every week. On this day. Every week. He was a good son.

'Call the boss man, Wei!' said Li, staring at the phone as if it could work magic.

Wei stabbed at the numbers and pressed the cell phone to his ear.

'You have reached the voice mail of…'

'He's not answering.'

'Leave a message.'

'Mister Boss. We're on sand. Sea comes in. Come and get us.'

Wei had worked hard to learn some English but his brain was numb. He couldn't say what he wanted to say. He just wanted to swear at the boss man in Chinese.

The sand bar was shrinking. Now the river round them had become a flood. A flood of frothing water that stretched all the way to the cliffs in the gloom.

One of the men grabbed Wei and looked hopefully into his face.

'Wei, you're a fisherman. You can swim. Go and get help.'

Wei shook his head. Fishermen didn't learn to swim. They learned to grab something and float. Those that tried to swim always perished.

'I can't swim. I'm sorry.'

'But you're a fisherman.'

'I can't swim. That's it.'

The man's hands slipped from Wei's collar and he staggered back to sit on his bag.

Li moved close and put his arm round Wei.

'What we going to do? Wei. We can't just stay here, can we?'

But Wei had no answer. They might get swept by the tide to the shore but there would be undertows – the currents that ripped in different directions under the surface. The chances of getting to shore were...Wei closed his eyes to the thought. He shook his head. The darkness was falling as the tide grew.

The whole journey from China seemed to be in darkness. From one truck to another. Day after day. Seeing nothing but the inside of a lorry, the faces of the other men, silent, bowed. Every once in a while they would stop, the doors would be flung open and they would stagger out into bright sunlight that burned their eyes. Always in the middle of nowhere – rocks and sand, a few withered trees. The men would relieve themselves or defecate behind the rocks, trying to crouch for privacy. The drivers would share out rice balls and sometimes chicken wings. No-one seemed to speak. Then back inside and the lorry jolted on, bumping and jarring them towards their final destination: In Glun.

Wei knew that if he could get through this everything would be all right. Everything. His future was bright. He tried to cheer up Li who was often sick by describing how things would be in that bright new future in In Glun.

One of the men – he came from the next village to Wei – had started crying and shaking. No matter what Wei said, Wei even hugged him, it made no difference. The man had lost the will to continue. At the next toilet stop he had tried to run off but the two drivers were after him. They were too quick and strong for the man. When they caught him, they threw him to the ground and began kicking him. His screams were pitiful. Wei ran forward and with his full weight bowled the first driver to the ground. But when he turned to face the other driver he was looking down the barrel of a pistol. No-one tried to run after that. The beaten man was thrown into the back of the truck and it sped off, scattering stones and dust.

After a couple of weeks in the truck, they had ended up on the coast somewhere. All the countryside was green, very green. The drivers handed them over to some white faces with small eyes who smiled and pointed across the sea.

'In Glun. You.' His fat finger pointed at them. 'You go to In Glun.'

There was a boat. It looked and smelled like a fishing boat. The men – there must have been forty or fifty of them now as other lorries arrived – crouched on the wet, slippery deck, their bags of belongings between their knees. The journey was at night and the boat showed no lights. Wei knew how dangerous this was – they would never be seen on the open sea. It weaved away from passing ships in case they were spotted. Wei was fine on the bucking boat – he could balance easily against the swelling waves. It was what he had done from a small boy. But the others suffered. Many were sick, throwing up over the side of the boat if you were lucky. The smell of vomit was everywhere. When they had finished vomiting and there was no more to come out, they coughed and groaned and held their stomachs in pain, folded double. There were bottles of water and Wei moved from one to another, pushing the bottles against their lips, making them sip, making them drink. Water that made them ill, cured them too. Wei knew that.

The boat came to a halt after hours of hell, a hundred metres from a dark shore. A light flashed a signal from the beach. It was the middle of the night and the men were ordered to scramble into the sea, up to their waists in icy, black water, holding their bags above their heads. On the beach, two men with hats pulled down and wearing rubber boots, hurried them off the sand along a narrow pathway. Wei heard the engine of the fishing boat roar sharply then it gently faded away. They walked, wet clothes flapping cold against their legs for what seemed an hour until they ended up at a farmhouse – no lights on, no sign of life. Their two guards ushered them inside.

The house smelled old and damp. A fine, grey layer of dust covered everything. The house had not been lived in for a long time. There were piles of blankets on a table that were shared out. A bag of old clothes was emptied on the floor: ragged jeans, pullovers, shirts and the men grabbed what might fit them. All the clothes seemed three sizes too big. One of the men in charge spoke some Chinese.

'Find a place to sleep. Tomorrow morning we bring breakfast and you meet the Gang Master. Then rest. Day after, work. Hard work. Lots of money, yes?'

He rubbed his thumb and finger together and smiled a broken tooth smile.

So this is In Glun, thought Wei, looking around at the drooping figures, tearing off their wet clothes and climbing into the dry ones. Li, holding his baggy jeans up with his hand, looked at Wei with tears in his eyes.

'It can only get better, can't it? Wei, my brother?'

Wei nodded and tried to smile. 'Of course.' But his heart was a dark place.

For Wei, exhausted, confused, and lonely, even with the company of Li, the first days passed in a haze. The Gang

Master explained they would work every day – some on a farm, some in the bay. For one month he would take their money to pay for the transport from China to In Glun. After that, he would take money for food and the accommodation. Wei's dream of making money to send home to his mother and father began to dry up. His fears were made darker when the Gang Master told them to keep silent away from the farm, not to talk to anyone and if the police did come by any chance, they knew no-one, especially the Gang Master. There was no Gang Master as far as he was concerned. But there was a pistol in his hand as he spoke. They would get one phone call home each month. In his other hand he held a mobile phone.

Wei kept his own mobile phone secret. He found an electric point in the barn that worked and poked the wires from his charger into the plug. He had left money with his parents to keep the phone topped up for six months. It had been nearly all his 'savings' but he knew they all needed to keep in touch.

Every morning two lorries arrived at the farm house. The drivers brought rice, vegetables, lumps of pork in a large pot. The same pot came at night when they returned. Always the same food. Wei, because he knew a few words of English, and because he was the biggest man, was told to be the 'captain' of the team. 'Man-You' the Gang Master called them. The Farm gang were 'Ars-Nul'. They were in In Glun. They were supposed to be happy. That was the last thing they felt.

Every morning, Wei was chilled to the bone. They had been given water proof tops and trousers for the cockle picking. But nothing kept the cold and the rain out. Wei remembered the hot winds that blew from the shore back home. How he wanted the sun on his bare back again. He thought of Ma and Pa, shading their eyes against the sun, gazing out across the sea, proudly watching their son working in In Glun. Rain mixed with Wei's tears. No-one noticed. Li had become more and more silent as the weeks passed and Wei could not rouse him from a gloomy sense of despair. His friend talked only of the cold, the wet, the misery of the work. That misery seemed catching among the men like a flu bug, spreading each day. Some talked of running away but then were reminded of Gang Master's gun. And where would they run to? They had no idea where they were. If they were found by the police they would be put in prison. One or two thought that would be better than working like this. Wei tried to encourage them to stay hopeful but his words sounded empty. He knew that. When they had enough money they could return to China. He dare not think how that might be done. Until then, they had work. And each other. He looked round at the thin faces, dirt lined, the greasy hair hanging down, the men stooped with tiredness. Yes, we will go back to China as heroes.

The sand bar was disappearing underwater beneath their feet. Every one of them was shouting, hoping someone would hear, get a boat and rescue them. All would be well.

But no-one came, because no-one heard. The wind from the cold, grey North sea whipped their voices away to nowhere. No-one bothered about the cockle pickers. They were invisible. They didn't exist. They were Gang Master's dirty secret. To others who might see them, they were crouching black figures in the distance. They arrived in the dark and left in the dark. They were nobodies.

The men huddled together, hoping they would save each other. Li looked up at his friend, desperately searching for an answer, something that would tell them what to do. He was the 'captain'. Wei always knew.

Some of the men swore loudly and flung their bags of cockles into the rising waters. Li was crying now, and Wei pulled him close, holding him out of the water.

Wei pulled out his phone for the last time, he knew. Pressed the buttons that would call home. The waves were up to their waists and it was hard to balance.

'Hello? Is that Wei?' His mother's voice. It sounded faint and crackly. 'Is everything good, son?'

A wave slapped against his chest. Li groaned and slipped from his grip, floating for a few moments, staring hopelessly up at Wei.

'Yes, Ma. Everything is good. The work is hard...'

Water splashed into his mouth and he coughed at its

bitter taste.

'Good, son. Good. We miss you.'

He stretched his neck, tilted his head back, holding the phone close to his ear. The water was at his throat.

'You are a good son, Wei.' His father's voice.

The next wave took him.

February 2004

23 Chinese workers, all from Fujian province, drowned.

They could earn HK$1,000 a week but owed $50,000 to the snakeheads who brought them to England.

Gang Master, Lin Liangren, was put in prison for fourteen years in the UK.

No compensation has been paid to the families of the victims.

Charities in the UK have raised a few hundred pounds (HK$3000) for each victim, just enough to cover funeral costs.

The profit from cockle picking in the area is HK$100,000,000 a year.

Thousands of Chinese illegal immigrants are still working in Eastern England in slave conditions.

IN GLUN. IN GLUN's SHAME.

Woofa the Dog

'Oh, look at the poor little thing.'

Mimi's first words when she spotted the puppy hidden between the two black plastic dustbins at the back of the block. The pup was a sad sight: its fur matted and greasy, the legs covered in cuts and sores, its mouth sagging open, the pink tongue lolling helplessly to one side.

'Be careful, Mimi. You don't know where it's been,' warned her mother as Mimi bent to encourage the animal out. On its stomach it crawled towards her open hand held towards it.

'It might be savage. It might bite,' came the voice of caution.

'Oh, Mother. Don't be stupid. Look! It's helpless.'

By this time the dog's rough tongue was brushing Mimi's palm clean and her other hand was rubbing the sticky fur on top of its head.

'And before you say anything, Mimi. No!'

'But Mum…'

'The flat is too small.'

'But Mum…'

'It will make a horrid mess.'

But Mum…'

'Just…no. No. No. I'm sorry, Mimi. No.'

'OH LOOK AT THE POOR, LITTLE THING.'

Woofa liked the kitchen (Woofa was Mimi's choice of name) because there was always food around and stuff might drop on the floor to be hovered up beneath the black shiny nose. It was Mimi's dog, without doubt. Neither her mother nor her father wanted it or liked it particularly but it kept the girl happy and seemed to give her something else to care about apart from herself. She was an only child after all. They'd wanted more children but it hadn't happened and Mimi consumed their time efficiently and without a break.

They smiled when Mimi cleaned the puppy up in the kitchen sink using half a bottle of washing-up liquid. Yes, it did look better after the Big Wash. Clean fur seemed to make the dog look bigger, didn't it? All the black fur fluffing out.

Mimi looked after Woofa completely and totally. She took him for walks on the end of a silver chain, everyday through the empty play area in the mornings and the squawking, clambering, shouting play area in the afternoon. The little children would smile as Woofa passed by, tugging on his chain to be released into the play area. But Mimi never let him go. Woofa was hers – not to be shared.

After a walk, Mimi carefully washed his paws, dabbing at the black pads with a soft, foaming cloth. She shampooed him every other day, using her hair dryer to fluff him dry. Oh, how Mimi cared for that pup!

'Who do you think he belonged to, Mimi?' asked her father, knowing the girl loved to talk about the dog.

'Oh, someone very important, I bet. You can tell Woofa has got…class. He must have just got lost one day and had to survive on rubbish and bits from the bins.'

'You don't think he was abandoned, then dumped? Maybe there's something wrong with him?' Her Dad was looking closely at Woofa as he spoke.

Woofa seemed to go tense suddenly and was staring at her father with gleaming, intent eyes. And did that doggy lip curl up a little to show a bright, white incisor – a fang?

'Something wrong, like what?'

'Oh, I don't know. Perhaps he was a bit vicious. Nasty. Some dogs are, you know.'

'Vicious? Nasty? Look at him, Dad!'

Woofa almost in response, had rolled onto his back and showed his fluffy black tummy to be rubbed.

'Woofa, you're not vicious, are you?'

Rub rub. Rub rub.

'You wouldn't hurt a fly, would you?'

Her Dad was still looking at Woofa with a curious look on his face.

'But no-one put up any notices, did they? You know: 'Pet dog lost'. 'Has anyone seen a black puppy?'

'No.' Could Mimi see where her father was going with this?

'I mean, if the dog is so cute – which you think he is – why didn't anyone come looking for him?'

'I don't know, Father. All I know is that he's cute and he's mine.'

Her father had moved closer, standing over both of them.

'Just seems a bit strange. If he was dumped, why was he dumped?'

He'd slipped his foot out of the slip-on shoe and was extending it towards Woofa's tummy. An accusing toe rather than an accusing finger. And that is when it happened. In a flash, Woofa's mouth had closed on the toe, the sharp little teeth clamping through the sock into the flesh and down to the bone.

'Aaaaaeeeegah!'

Her father's scream of pain filled the room, rattling the windows, burning into Mimi's brain.

'Get it off!'

He tried to bend down but the puppy moved back out of reach, tugging at the toe, tearing it. Blood was welling through the sock.

'Get it off me, Mimi!'

Mimi flung herself forward and grabbed Woofa.

'Let go, Woofa! Let go!'

But Woofa would have none of it. He tightened his grip on the toe, encouraged by the father's squeals of anguish.

'Woofa! Let go! Let go!'

Mimi's fingers prised at the puppy's jaws, trying to open them and release the toe, now pulsating with blood. But the jaw bone was rock solid. How could the pup have so much strength? She clasped the two hinges of the mouth and tried to pull them apart but…nothing.

Her father now took to trying to hop away from the pup but merely dragged the pup with him. His toe was shredding rapidly.

Woofa's black gleaming eyes fixed on Mimi and she could see hatred and anger burning there.

'Woofa! He didn't mean it! He didn't mean what he said. Say you're sorry, Dad!'

'Aaaagghh! I'm not saying sorry to that wretched puppy! Just get it off! Aaaaagh!'

Woofa's grip tightened even more and her father thought he was going to lose his toe, that the sharp little teeth would go all the way through the bone.

'Just say it, Dad! Just say it!'

In a strangulated gasp, her father muttered, 'I'm sorry, Woofa. I'm sorry.'

Instantly the animal released its grip, rolled onto its back and smiled up at Mimi even though its mouth was flecked with her father's blood.

Her father fell back onto the settee, holding his toe as tightly as possible to stop the bleeding. He looked with horror at the reclining puppy and then at his daughter who was gurgling apologies all over the animal.

'Don't sit there rubbing its belly. Get me some plasters and bandages!'

Mimi lifted the dog into her arms and marched out to the kitchen, depositing Woofa carefully and deliberately into his basket. A few moments later she was tending her father with his shredded toe and his shredded sock.

'That dog is a menace, Mimi!' he hissed through the pain as the bandages tightened round his toe.

A sharp bark came from the kitchen and her father stopped talking, concentrating on putting his toe back together. And so the incident passed.

Mimi thought a lot about what had happened. She wasn't worried about her father's toe but more about how Woofa seemed to know what they were talking about. Surely the dog couldn't understand human communication? Words maybe. Sit! Lie Down! Be quiet! But accusations? Bad-mouthing? Surely not. Or was it just the tone in the voice?

Woofa was absolutely normal after that, playing and rolling around, as puppies do. Strutting through the playground, soaking up the other children's 'oooh's' and 'aaahh's'. Mimi's Dad kept his distance, aware that Woofa was watching his toe, a little snowman on the end of his foot, with interest. But there were no more attacks. Mimi's mother blamed her husband for upsetting the dog. The father's shoulders slumped and he eyed the dog menacingly. Woofa slowly wagged his tail, looking at him all the time.

'Woofa does have an enormous appetite, doesn't he, Mimi?' complained her mother a few weeks later. There had been no more attacks on toes or fingers or anything – just a puppy with a ravenous appetite. Every other day, Mimi was staggering back from the supermarket dragging a huge bag of dog food: tins of meat, biscuits and bones. And away it would all go – woofed by Woofa. The food went in but Woofa didn't grow, didn't get fat, didn't burn it off with lots of energy. Mimi noticed he didn't even poo much. Not like other dogs she saw in the park, leaving their steaming deposits everywhere. So where did the food go?

'You're a greedy little pup, aren't you, Woofa?' smiled Mimi holding the dog's nose against hers. Woofa burped a doggy-smelling burp all over her face.

Mimi knew something had happened as soon as she saw Woofa in the morning. He looked…a bit exhausted, spread out on his side, stretched out, tongue lolling. She knelt beside him.

'Where have you been? What have you been up to?'

Woofa rolled his eyes and looked away.

'The door was open this morning, Mimi. Did you let Woofa out in the night?'

Mimi looked up, frowning to match the serious look on her father's face.

'No. Of course not. Someone else must have, Dad.'

'And who would that someone else be? Your mother? She

didn't. And I didn't.'

'Well, Woofa couldn't open the door, could he? He's a dog.'

Woofa purred like a cat. Mother, who had been holding back, stepped forward.

'Maybe we made a mistake. Perhaps the door wasn't shut properly.'

Mimi nodded quickly.

'That's most likely it.'

Her parents exchanged a look that Mimi had trouble trying to interpret. Did they think she was lying?

Mimi was leaving for school, rubbing a sleepy Woofa's tummy as she always did before she left. As she got to the front door and opened it, she jumped. A policeman with a notebook and a serious face was standing in the doorway, his finger poised to ring the bell.

'Excuse me, Miss. Is your father or mother in?'

Mimi revolved her head and shouted for her mother.

'Mother! It's a policeman at the door.'

A brief second passed and her mother was by her side, gazing wide-eyed at the dark blue person in front of her.

'Can I help you, officer?'

The officer looked down at his notes.

'I've received complaints from other people in the block about someone running around all night, causing mischief in the lifts and the stairwells. Hours of noise and running around apparently. We think it may have been burglars – probably illegal immigrants – trying to break into the flats. Did you hear anything? See anything?'

Mother was staring so hard at the man's face that she forgot to answer.

'Well, did you?' repeated the officer of the law.

'No. Nothing.' Mother shook herself. 'Except…yes, our door was open this morning. Most unusual.'

'And has anything been taken?'

Mother scratched her head absently. 'No. I don't think so. The dog would have barked.'

Woofa had roused himself, padded into the hallway and sat looking up at the policeman as if he wanted to join in the conversation. The policeman nodded and folded his notebook away in a pocket on his chest.

'Well, make sure you lock the door tonight and listen out for troublemakers.'

'We will,' nodded mother. Woofa barked once, sharply.

The policeman tipped his hat and backed out of the doorway. Mother clicked the door to close and looked enquiringly at Mimi.

'Well. What do you make of that?'

Woofa settled back to his basket in the kitchen. His tail wagged confidently.

That night, Mimi lay awake in bed unable to sleep, listening out for noises in the corridor. There were none. Eventually sleep overwhelmed her and she woke suddenly with a bitter taste in her mouth, with the sun shining through the curtains and with a fuzzy head. Her mother was standing over her, tense and intense.

'The door. It's happened again and I know I locked it last night. Your father checked it too.'

The clouds in Mimi's mind cleared in a second.

'Perhaps it's a faulty lock.'

'No. I don't think so. And there's something else you need to see.'

Alarm bells buzzed in Mimi's head and a feeling of dread filled the pit of her stomach. She followed her mother into the kitchen and looked down at where an accusing finger was pointing.

'Look.'

Woofa was lying curled up, fast asleep, breathing deeply, one leg kicking occasionally as doggy dreams flitted in and out of his doggy brain.

'Look!' her mother repeated more urgently. Then Mimi saw it. Woofa. His fur was ragged and sticky, hanging in clumps, greasy and sticky and tangled. Mimi bent down and tenderly touched it, smelt it. It was like iron filings – an almost bitter smell. Then she saw it around Woofa's mouth, on his teeth, as his mouth snuffled open. Blood. Slowly Mimi rose to her feet and looked at her mother.

'He's been out, hasn't he?' whispered her mother. 'He's been out and he's…killed something.'

'Killed something? What? What could he kill in the dead

of night?' gasped Mimi.

'I don't know,' said her mother. 'Cats, rats, another dog…'

'But he's a puppy. He's only young. He doesn't know how.'

'He knew how to kill your father's toe.'

Mimi stepped away and turned quickly.

'That couldn't be. How could Woofa open the door? How could he kill something? He's too young. He's too small.'

Mimi turned her daughter around.

'Something is happening and it's not very nice. I don't know how or why but it's something to do with Woofa. Now clean him up before you go to school. And don't say anything to your father. Not yet.'

'Why not, Mother?'

'He's taken against the dog since the toe incident and he'll want to get rid of it. I know you don't want that.'

Mimi knelt down towards Woofa who, almost magically, opened his eyes, wagged his tail and tried to lick Mimi's nose.

'He couldn't have done any of those things, Mother. Just look at him.'

Her mother slowly turned and walked away, shaking her head.

Mimi was surprised at how much blood there was on the dog's coat. It turned the sink, full of water, red, three times over. She didn't have time to clean the blood from around his mouth. She rushed off to school, leaving a damp puppy to which she, for the first time, had not said goodbye.

At school, Mimi immersed herself in her studies, not wanting to talk to anyone. But at break time, it became obvious something 'big' had happened. Some younger girls were in tears, some being comforted by teachers, others sitting in tight groups, hands together. Mimi moved towards one group – she knew several of them lived on the same estate as her.

'What's happened? What's wrong?'

The Year Two girl – Sha Zi – Mimi knew her – looked up, eyes red with tears.

'My dog's been killed. Last night. Had its throat torn out.'

Mimi gulped and staggered back.

'What?'

'And that's not all. Four or five others here – the same thing. Pets butchered. A pet rabbit. A cat. Another dog. It's awful.'

'It is awful,' repeated Mimi mindlessly.

'Something's on the loose. A pet killer.'

Mimi looked down at her hands. Were they slightly pink from the blood stains this morning? She wiped them hard on her skirt, hard, several times.

There was still nearly a whole day to go. Maybe she should go home. Check on Woofa. Would Woofa be out again while she was away, doing horrible things. She banged her head with her fists. She had brought these problems on when she took in the puppy. It was her fault. She was to blame for all this.

There was a note from her mother, on the table when she got home thirty minutes later.

'Mimi. The police called again. Other pets have been killed. They're looking for the killer. Don't go anywhere till I get back. Mum.'

Woofa padded in and stuck a damp nose on her knee, looking up hopefully at her. For love? For food? For something else?

'What's going on, Woofa? What have you been up to?'

Woofa just rolled on his back, paws in the air. How could this animal do all those things to other animals? Not this one. Not Woofa.

The policeman called again – this time there were two of them. Mimi's father hadn't got home from work. What would he do this time if he found out the circumstances? Woofa would go. Maybe handed over to the police. And then what? Locked away or even...

Mimi couldn't contemplate that. Mimi's mother was standing at the window staring out.

'How does he do it, Mimi? How does he get out? Are you sure you don't let him out? Perhaps by mistake? Sleepwalking?'

Mimi glanced into the kitchen. Woofa was in his basket, head resting on his paws, looking at her. Mimi went to the kitchen door, pulled it closed then went to her mother.

'I don't know, Mother,' she whispered.

'Why are you whispering?'

Mimi reached up and put her hand across her mother's mouth.

'I'm whispering because I think Woofa is listening to what we say.'

'What? You mean it understands us?'

'Yes.'

'What sort of crazy dog is this that you've brought here, Mimi? If it can understand us, it can probably open the door. It's worked out a way to do it.'

'I don't know, Mother. I don't know. What shall we…'

At that moment the kitchen door clicked open and swung wide. Mimi and her mother squeaked with fear and jumped around. Standing in the doorway was Mimi's father. There was a long, painful silence.

'Isn't anyone going to say hello?'

From the kitchen Woofa barked. Mimi and her mother stared at each other.

'Hello, Father! Good you're back!'

Her father stared at her questioningly.

'I'll make a cup of tea, Father,' smiled her mother, going to him and urging him into the living room. When she went into the kitchen, she stepped warily around Woofa.

Mimi fiddled with her fingers as her father sat down and looked around.

'Well, anything happen tod…'

'No,' cut in Mimi too quickly. She stared at him while

words tumbled around between her brain and her mouth. Then she dashed towards the door.

'I'll help Mother with the tea!'

Her father smiled.

'Well, this is new. You, helping in the kitchen.'

Inside the kitchen, Mimi pressed the door shut and leaned against it. Her eyes were wide, taking in her mother poised over a tea pot and Woofa's eyes watching her.

'What do we do? Tell him? Tell him something different? Make something up?'

Mimi's voice sounded like flints – on the edge of desperation. Her mother put the tea pot down and moved to stand close to her daughter. She whispered in Mimi's ear.

'No. Say nothing. Tonight, we'll keep guard on…our friend.'

She ventured a look at Woofa who was inching closer from his dog-bed. Mimi closed her eyes and let out a long sigh.

'Okay. Let's hope it works.'

Mother had managed to persuade her husband to go to bed early. She then busied in the kitchen, apparently preparing the next day's meals. The plan was she would stay in the kitchen with Woofa till midnight. Mimi would take over from midnight to two, her mother would see the rest of the night through.

'Good plan,' agreed Mimi, blocking out the thought of how they could continue guard duty for more than a couple of days.

And all seemed to go to plan. The two flitted silently through the early hours while Woofa slept, head on paws.

Mimi woke with a start. She had been dreaming horrible dreams of dogs attacking humans, eating tower blocks, dogs ruling the world, marching humans to huge prison camps. She was in a cold sweat when she woke. The clock said six.

She rushed across the room to the kitchen and there was her mother sat at the table, her head on her arms on the table top, fast asleep. She turned to look at the dog basket and there was Woofa, fast asleep, covered in filth and gore and smelling of...well, death.

Mimi shook her mother awake.

'Mother! Mother! You fell asleep. It's happened again!'

Mother struggled to wake up, rubbing her eyes, holding her head, her whole being thick with sleep, with unconsciousness.

'What? What happened?'

Mimi looked along the hallway and saw the front door was ajar.

'It happened again. Woofa got out. Look at him!'

Mother's face dissolved into agony when she saw the animal curled up in its bed.

'Oh no, I must have gone to...no, hold on. I didn't. I remember. It was three o'clock. Woofa woke up and began staring at me. He waved his tail slowly from side to side and ...I fell asleep. In a trance.'

Mimi gulped. 'He mesmerised you.'

A voice came from her parents' bedroom.

'Everything alright? I can hear you chatting!'

Mimi and her mother fixed gazes on each other.

'Clean him up! Now!' hissed the parent.

Mimi swung the somnolent dog upwards and straight into the kitchen sink. Red splashed onto her arms as she turned the taps. The dog was half asleep still, floppy and unresisting, as the hot water and the scrubbing brush Mimi held, scoured the awful stuff from its fur. Mimi didn't bother with the plug, letting the disgusting brown and red liquid flow away. When she'd finished, she took a tea towel and rubbed the animal hard all over then dumped him in his dog bed.

'Stay there, Woofa. Don't move.'

But clearly Woofa wasn't going anywhere. He settled, curled and went to sleep.

'Oh, this is hopeless,' breathed Mimi. 'Absolutely hopeless.'

Her mother's advice – no, instruction – was clear. They were talking in low voices outside the closed front door.

'We can't go to the police now. We should have gone days ago. We'll be arrested for…with-holding information or something. Mimi, you go to school. Act normal. Say nothing. I will sort something for tonight.'

Mimi was dreading going to school, to hear a barrage of stories of mutilated pets. All murdered – yes, murdered! – by her own puppy. Woofa. Except there were no stories. There was a police investigation going on but why hadn't they come to Mimi's apartment? The panic that welled inside Mimi gradually settled. Until suddenly the bigger question popped up in her head. Then what had Woofa attacked during the night?

There were no answers, no clues until that evening. Her mother was at the front door when Mimi reached for her key to let herself in. Her mother pulled the door shut and once more they spoke in low voices in the corridor.

'Mimi. The police phoned.'

'I knew they were investigating the pet killings.'

'Worse than that.'

'What?' Mimi felt as if cold slime had been poured down her spine.

'Look at this.'

Her mother led her inside the apartment to the living room where the television was burbling away – the national news. A man in a suit and tie was staring seriously at the camera.

'…the attack came in the early hours. The body was found in a rain channel in Promise Gardens. It had been badly mutilated. Police say this was a savage attack, possibly by a maniac or even a wild animal like a wild boar…'

Mimi grabbed the remote and the screen went blank.

'Promise Gardens! That's at the end of the street! Mother!'

Her mother put her finger across her lips and closed the

kitchen door. Woofa was in the middle of a major, slurping intake of water. The two spoke in urgent whispers. Mimi's eyes were gaping wide.

'That was Woofa, wasn't it? He's killed a person now. What are we going to do?'

'Mimi, can we be sure it was the dog? It could be a person – an ordinary murderer.'

'What does an ordinary murderer look like, Mother? I know what an extra-ordinary murderer might look like. Small. Black. Four legs. Fur. Goes by the name of Woofa.'

'We can't be sure. And we can't go to the police and report Woofa. Not now. And who would believe a puppy could hypnotise a human, open a locked door, get out of a block of flats and kill a person in the street. They'd lock us up for wasting police time.'

Mimi looked desperately around as if the solution might be written on the wall in poster paint.

'What do we do, then? I've washed away the proof – the evidence. Woofa is just a little puppy again. Except he needs feeding again.'

Was that a little whimper they heard from the kitchen? Was the animal able to hear through doors and walls? It had an incredible appetite. Perhaps its senses were incredible too.

'First thing.' Her mother was tapping her teeth. How did she stay so calm in the middle of this chaos?

'We don't tell Father. If we did, he would just take Woofa away and…do something final.'

'If Woofa let him. This puppy has already killed a man. Remember that?'

'Well, we don't know for sure he did. But, good enough reason for not telling Father.'

'So what do we do?'

'We have to fool Woofa. Let him get out, follow him, see what happens …then…take action.'

'Take action? What does that mean?'

'We have to do what we have to do.'

'You mean, ki…'

'I mean, take action. In the meantime, we must act absolutely normally in front of Woofa. He will sense – he will know – if we are not absolutely careful. Right?'

A long pause.

'Right.'

Mimi thought her mother must be a spy or a gangster or a secret detective, the way she thought things though, prepared. Mimi went back to the kitchen, knelt by the dog basket and began tickling Woofa's tummy. He raised a smiling head and licked her hand. Did the dog notice how her hand jumped as she thought of the human tissue that had been in the dog's mouth? How could she now love this…thing, this monster that was locked inside her life?

'Oh, Woofa. I have to go to school. You stay here. I'll see you later.'

She gave his head a fleeting rub, picked up her school bag and marched quickly through the front door, shouting

'Bye Mother, bye Father' as she left. She was way too early, hadn't had any breakfast and her brain was whirring with anxiety. She was dreading school. Dreading home. Dreading the night to come.

At school assembly, gathered in the heat in the large courtyard, a police officer in uniform stood next to the Principal who hovered uncertainly by the microphone.

'Girls, I have a special visitor here today and I want you to take notice of what she has to say. It is very important.'

Mimi felt a wave of nausea sweep over her. She wanted to vomit. The police were onto a murderer and had traced them to the school. It would only be a matter of time before she was arrested for aiding and abetting a murder. They would have DNA. Her DNA at the crime scene. She shuffled behind a taller girl out of the sight line of the police officer in case she could see her guilt from thirty metres.

'Girls, I come with a warning and a request.' The voice tone was sharp and accusing to Mimi's ears. 'The night before last someone was murdered in the Promise Gardens. We believe that whoever killed the victim was not human. We are talking about a wild animal – maybe a wild boar or even a wild dog or pack of dogs. This animal is out there somewhere and armed officers are out tracking it. Until it is captured and destroyed, we suggest you do not go anywhere within a mile radius of Promise Gardens as you will run the risk of being attacked. Until we find this creature it will be better to stay indoors at night for the time being. For your own sakes. Thank you.'

Mimi shuddered. She was living with this creature. She let it lick her hand. Her nose. She filed back to the classroom somehow managing to put one foot in front of the other, not talking to the others who were excitedly discussing what sort of animal it might be that could kill someone. A rabid monkey? A vicious snake? Mimi wanted to shout out: 'What about a cute little puppy?' But she didn't. She just felt miserable with the uncertainty of knowing what to do. The police would never believe them. It seemed impossible to control Woofa themselves. He had control over them. It was only a matter of time before he… she blanked her mind to the ideas of more murder and mayhem.

On the way home she called in at the supermarket for her daily haul of pet food. Filling the basket, she wandered the alleyways of goods as the basket got heavier. Then she saw it in the specialist aisle. A possible answer. She grabbed two red boxes from the shelf and stuffed them in among the dog biscuits and tins of meat. What she now needed was something really tasty for Woofa.

'You got what?' her mother exclaimed, looking at the basket perched on Mimi's bed.

'Keep your voice down, Mother. I bought some rat poison and some lean pork. Woofa loves pork. It says…' she read from the red box,' soak your bait in the poison for three hours then blah blah blah…'

Her mother looked down at her feet. 'It will be a very painful way to die. You do know that, don't you? Woofa will go mad with the pain.'

Tears welled in Mimi's eyes. She didn't want this. She had brought the dog into the house. She had loved it so much. She still did, despite everything. And yet here she was planning to murder the object of her love. She had to solve the problem herself. She had to put her feelings to one side.

'What else can we do, Mother? What else can we do?'

Mother shook her head. 'I know, Mimi. We must do this together. We mustn't involve Father. He will go mad himself if he finds out and blame us for what has happened. He's working late tonight so we can prepare everything.'

Mimi didn't dare do the mixing of the poison in front of Woofa in the kitchen, just in case it understood what was going on. Smelling poison? Reading the instruction box? Mimi cursed her own stupidity at thinking that. She cooked the pork in a pan with Woofa looking up at her expectantly. If the dog had looked closely, it would have seen spangles of tears in its mistress's eyes.

'Oh, Woofa,' she muttered. 'Oh, Woofa. This is a treat for you later. The treat of a lifetime.' She swallowed the treacherous words.

Soon the pieces of pork were suitably laced with the poison which had no smell (according to the red box). Mimi washed her hands carefully and put the plate in the fridge.

Then she decided she must take Woofa out. He was used to a walk when she was home. Perhaps no walk would make him suspicious. Mimi kicked herself again for thinking the dog could be human. But then...

Woofa enjoyed the walk. He behaved like a puppy should, sniffing at everything, running forwards as the lead played out, scampering back when another dog came along. Woofa even marked out his territory, peeing quickly here and there. What if they were wrong, thought Mimi and Woofa wasn't doing all these things. That he was just a puppy. Innocent? And here they were planning to kill him off. This little bundle of black fur? Mimi was almost lost in these notions of the animal's innocence until they got to the street that led to Promise Gardens. Suddenly Woofa was pulling hard to go towards the green wood and the fountain that made up the gardens.

'No, Woofa! This way!'

The dog strained at the lead and Mimi was aware of the enormous strength that this little furry body contained. She staggered under the fierce tug. Resist as she might, she could only but move one step at a time towards the gardens, pulled relentlessly by the black ball of energy. The dog was impossible to control. She knew that.

And then Woofa stopped. Stock still. Looking ahead at the Gardens, Mimi saw what the dog had spotted. Across the entrance was brightly coloured tape with 'Police Crime Scene' and in the gardens themselves were two people in white suits moving about.

Woofa turned and, wagging his tail like any ordinary pup, tugged to go home. Dread filled Mimi to the brim. What sort of dog was this that could recognise the police in that way? A dog that could do all this? It had to be the Devil's dog, surely?

Mimi told her mother of what had happened. Her mother shook her head.

'We have to be careful here, Mimi. So careful.'

They were whispering to each other in the toilet, pressed against the toilet paper holder.

'But how are we going to keep Woofa in tonight? We don't know how long the poison will take.'

Mimi couldn't quite believe what she was saying. The thought of deliberately killing a dog appalled her but here she was doing just that.

Her mother patted her hand. 'I have a plan. I have a little bottle of sleeping tablets – I haven't used them for ages. Three or four of them will knock him out. I'll dissolve them in the tinned meat. Don't worry, Mimi. We'll sort it.'

She could see the tears welling in her daughter's eyes and she pulled the girl to her, wrapping her arms around her tightly. So the best laid plans were made. They would behave perfectly normally. The poisoned meat they would give to Woofa at the same time as the sleeping tablets – about midnight when Father had gone to bed. By morning perhaps their troubles would be over. The story for Father would be that the dog had just died – perhaps choked on some food.

Mimi couldn't concentrate on her homework. The words and questions were a blur. It was all so pointless. She closed the books and closed her eyes.

Mimi woke with a start. She'd fallen asleep over her books. The strain of all this dog business had exhausted her. She looked at the clock. Eleven thirty. She padded out into the living room to find her mother fast asleep. There was a shuffling sound in the kitchen. Father had returned home.

She stepped into the kitchen – the bright light made her eyes ache. But she could take in the scene. Woofa was in his basket, bright eyes watching everything as usual, his tail automatically wagging when Mimi appeared. Her father was standing at the table, his body bent wearily. A weak smile tugged at his lips.

'Hello, Mimi? What are you doing still up? I've been working late. Mother still asleep? I didn't want to wake her. Are you alright?'

Mimi had gone very pale and had slumped against the door. Her brow and lips were waxy with sweat. Her eyes were fixed on the plate by her father's hand. The empty plate.

'That pork was delicious. Some unusual spice in it. Didn't expect to find that. I was feeling really hungry.'

The words crawled slowly out of Mimi's mouth.

'You ate the pork.'

She was aware of her mother now standing behind her.

'Mother. He ate the pork. Father ate the pork.'

Mother pushed past her.

'Father! That was for the dog! It has some anti-worming medicine in it which will make you ill. Go to the toilet now!'

Mimi watched her mother go into action. Mimi muttered, 'Respect!'

Her father was firmly guided to the toilet, made to kneel over the bowel while his wife slid two fingers down his throat. Mimi looked away as her father started gagging, his back rising in spasms, then the coughing, the coughing, the splashing, the moaning, the mother's hand patting his back tenderly, the final cough and her father straightening at last. Mother pushed the toilet lever decisively and the awfulness was washed away in a roar of water.

Woofa had come to watch, pressing against Mimi's leg. Father rose unsteadily to his feet.

'That'll teach me to be greedy. Oh dear, I feel a bit faint. I'd better get to bed. I feel so tired.'

'Good idea,' nodded his wife. 'Drink some water before you go, just to flush it out.'

With Father despatched to the land of sleep, the plotters met again in Mimi's room, speaking in hissing whispers.

'That ruined it! What are we going to do?' gasped Mimi.

Her mother held her shoulders and looked into the girl's eyes.

'We go on with the plan. I want you in the kitchen at one o'clock.'

'But what…?'

'Just do as I say.'

There was a cold, hard tone in her mother's voice which she couldn't argue with. Mimi slumped onto her bed and stared at the ceiling. Everything seemed to be crazy. It all seemed so wrong and yet she was carried along in this tide of wrongness.

One o'clock. It was pitch black in her room. She could hear rain splattering down against the window. She padded into the kitchen where her mother was busying, working in the dim light that came from a lamp she'd placed on the table.

'What are you doing, Mother?' she whispered.

'What does it look like?'

Then Mimi saw clearly what was happening. Her mother had a soft zip bag, opened on the table. And into the bag she was sliding the bundle of black fur that was Woofa.

'I gave him some extra tablets which should knock him out for a few hours.'

Mimi stared at her mother.

'So where are you taking him?'

Her mother paused, her hands resting on the animal, rising up and down with his breathing.

'To the Lantau Bridge. We are taking him to the Lantau Bridge.'

And then Mimi saw it all. The cruelty. The necessary cruelty. The bag turning end over end in the air, dropping. Dropping. The final splash. The bag disappearing into the dark waters.

Her mother's face was hard as she clenched her jaw.

'I have another bag – all the heavy tins of food. We'll add it to this bag to make sure it sinks.'

Mimi opened her mouth to speak but her mother cut in.

'Mimi. It's the only way. It's the only way to stop all this terrible stuff happening. You don't have to come. I'll manage it on my own.'

But Mimi knew she had to see this through. She had brought this on the family when she'd insisted in bringing Woofa home. It was horrible but it had to be done. What was the alternative? What might this crazy dog do next? More murder? More killing? Who would be the next victim? And what would happen if the police traced the killer back to them? They would be on trial. A notorious case – in all the papers, on television. Their lives would never be the same. Wave goodbye to your future, Mimi. No university. No nothing. The dark clouds rolled into her brain like a gathering storm.

Twenty minutes later saw the two sitting silent and white-faced in the back of a red taxi heading for Lantau Island. The small hold-alls were in the boot. Mimi and her mother stared straight ahead, listening to the purr of the engine, the click of the windscreen wipers as they scooped water from the screen, the ching! of the meter as each kilometre was ticked off the journey. They watched how quickly the raindrops replaced each other, how the lights shone red and gold and green like jewels, how the whole world outside seemed distorted through the rain-spattered glass.

The taxi driver tried to start a conversation, perhaps wondering why these two might be travelling like this, at this time of night. Running away, perhaps? A bullying father? It happened a lot. A secret meeting? Mother made it clear she didn't want to talk.

Ahead the high struts and metal towers of the bridge appeared white in the night sky. Mimi looked at her mother who stared straight ahead. Was that the sound of something moving in the boot of the car? Had Woofa woken up and started panicking? Would he break out? Mimi knew how incredibly strong he was. He could bite through the bag. When they opened the boot would he leap out and tear out the taxi driver's throat? Or even theirs? Mimi's legs were trembling till her mother's hand clamped on them.

'Stop it!' she hissed. 'Control yourself!'

The car was swinging off the dual carriageway under a motorway sign that indicated 'Lantau/Airport'.

'Perhaps we should put Woofa on a flight to China,' whispered Mimi then regretted it. Her mother's scornful eye burnt a hole in her brain.

'This is no joke, Mimi. Shut up.'

Her mother leaned forward and spoke to the driver over his shoulder.

'Pull over, driver. That lay-by. Li do.'

The driver half looked round with a doubtful expression on his face.

'You're miles from anywhere. It's very wet out there.'

'Just pull over. We're fine.'

He shrugged. 'You're the boss.'

The red car slid to a halt. Mother glanced at the meter and passed over a couple of notes.

'Keep the change.'

'Oh, thank you,' mumbled the driver, folding the notes and sliding them into his top pocket.

The boot clicked open. The two females stepped out into the rain, shuffled to the back of the car and lifted out the two bags. Mother raised her hand as the taxi pulled away. Then she pulled out two dark woolly hats – beanies – ones they used in the winter.

'Pull it down low – there may be security cameras.'

Mimi tugged the edges of the hat down beneath her ears and low over her eyes. Her mother really was a spy or a criminal. Or maybe she had seen too many Hollywood films. In their hats they both looked very suspicious thought Mimi. People doing wrong things dressed like this.

There was no path from the lay-by. They walked along the edge of the carriageway. An occasional car swished by, spraying them with water. One car angrily hooted his horn at the two figures and their burdens heavy on their backs.

At the bridge, her mother pointed to a narrow metal track that ran between the roadway and the edge of the bridge. A sign said 'No Public Access'.

'Come on, Mimi,' urged her mother as her daughter hesitated.

They walked in single file. The pathway had filled with puddles and Mimi's feet were soaking. The rain stung her face, driven as it was by the wind here on the heights of the bridge. She didn't want to look down – the dizzying height would make her feel sick, she knew. Her bag – filled with the tins – was heavy and she had to ease the bag continually between her hands. Her mother, ahead, carried her bag as carefully as possible, not wanting to rouse its contents.

They had gone about two hundred metres along the bridge. They had ducked low as any vehicle came along in case someone saw them, stopped to see if they could help. There was a lull in the traffic. The headlights no longer swept over them. Just the sound of the rain pattering all around them.

'Right, Mimi. This will do.'

Her mother placed her bag gently on the floor and gingerly unzipped it, half-expecting a black bundle of deadly energy and danger to come bursting out. But the pills had stilled the crazy.

Mimi placed her bag alongside the other and her mother quickly moved the tins of food and placed them around

'SAY GOODBYE TO WOOFA, MIMI.'

Woofa's blackness.

'Say goodbye to Woofa, Mimi.'

Her mother stood then moved several steps away, looking out over the black water, hundreds of feet below. Tears welled in Mimi's eyes as she knelt and reached out to Woofa for the last time. Her fingers sifted tenderly through his black fur, found his floppy ears, his damp nose.

'It shouldn't be like this, Woofa. You shouldn't have done those things. Forgive me. I have to do this. To save us. Goodbye, Woofa.'

A doubt gnawed at her throat, stopping her breath. What if they'd been wrong? What if Woofa had done none of these things? They'd just imagined it all? She bent over and delivered a kiss onto his bony head.

'Okay, Mimi. That's it. Well done.'

Mother pulled the zip across. The one holdall was now much heavier and it needed to be lifted to head height to get it over the parapet of the safety barrier. The two struggled to raise it, so afraid of dropping it and waking the animal. Grunting and straining, they finally balanced the bag on the rail.

'This is it, Mimi. Goodbye, Woofa.'

And with a slight push the bag disappeared into the blackness. They heard nothing. Woofa had become part of the blackness. Mimi felt her knees go to jelly and she began falling but her mother caught her and pulled her close. No words were exchanged. Both were sobbing now, great sobs that shook their bodies locked together. Rain and tears melted together and formed pools around their feet. A tug somewhere below let out a mournful hoot.

'It's over, Mimi. It's over. Let's go home.'

But what did that mean now to Mimi? A place of safety and love? Wouldn't Woofa haunt their home forever? He'd made it his territory. How would they be able to take it back for themselves? Everything had been changed utterly. Nothing would be the same anymore. Nothing. Mimi raised her face to the pouring rain and howled helplessly like a hound.

Hundreds of feet below, a launch was churning home through the dark waters. Young Jonny Li was sitting on the deck, rubbing the rain from his eyes. He was staring at the water fifty yards away. There had been a big splash, something dropping from the bridge. It could have been a person. A jumper perhaps. He stared up at the huge edifice of engineering, cold and white in its own lights. He stood up.

'To! To!' he called to the driver. 'Pull up! Pull up! Something dropped from the bridge!'

He hurried to stand alongside To, grabbed one of the searchlights mounted on the little boat's helm and turned it in the direction of the splash. Something big was just about floating but going down bit by bit. It could be a body, he thought.

'Let's pick it up,' Jonny urged.

The boat lurched as To span the wheel. As they drew alongside, the engine was cut. Waves slapped noisily against its side. Jonny grabbed a long hook from the bottom of the boat, leaned out and snared the bag even as it was sinking. He dragged it to the side of the boat and, with a huge effort that swelled his muscular arms, pulled the holdall onto the deck.

'Open it up then,' said To, imagining perhaps lots of money. Well, it happened in the movies didn't it?

Jonny lifted the wet, black bundle from the bag and held it up in front of him. He looked with a sneer at the heavy tins in the bag.

'Look at that. Someone's tried to kill this little dog, the heartless swine.'

At that moment the black bundle stirred.

'He's still alive! He survived the drop!'

Jonny's smile was wide. He'd always fancied a pet dog and had never got round to organising it. And now here was one delivered to him. It was fate. He imagined running through the park with the dog bounding along beside him. The dog greeting him each night when he got home. Suddenly this dreary night had become brighter. It really was fate!

The little dog's black eyes were now looking up into Jonny's eyes. The animal scrambled with his paws and reached forward to lick Jonny's fingers. The rough tongue tickled. Jonny laughed. To laughed at Jonny laughing.

'Will you look at that! He loves you already, Jonny.'

Thirteen Candles

Coulrophobia: the fear of clowns

It was Busy's thirteenth birthday party. She wanted all her school friends round of course. Her mother would buy a big chocolate cake and there would be thirteen candles. 'Thirteen?' someone asked. 'Isn't that supposed to be unlucky for some?'

'Only for gweilos,' said Busy. 'It's number four for us.' And she'd smiled confidently.

The last few days before the Big Day, she'd felt a bit nervous, excited, happy – a whole lot of things. Maybe saying goodbye to being a child? She looked at herself in the mirror. She didn't feel any different from being twelve. Just thinking differently. Expecting…more. Something exciting.

That's when the advert in the China Post caught her eye.

PARTY TIME?

GET A CLOWN!

Laughs, tricks, songs, juggling, magic.

All ages.

Mister Mayhem 152-666-999

Now that would be funny. If Mister Mayhem was rubbish, they'd spend all the time laughing at him. If he was good… well, that was cool. Busy's father agreed when she asked.

'Why not? It'll be fun. I'll book him.'

'No, Dad,' Busy insisted. 'My party. I'll do the booking.'

This is what thirteen year olds did. Took charge of their lives. And she did. Mister Mayhem sounded cute on the phone. He made a joke out of her name – something about being Lizzie too. She didn't get it. But, yes, he'd be there. And they were young teens?

'That's fine. I have just the right thing for you.'

She'd put down the phone before she remembered to ask if he wore a clown outfit with a clown face. Maybe he would be a modern clown and not do all that old stuff. Maybe just some mime of trying to get through windows and glass doors. Uncool.

The day arrived. Busy opened her presents at the breakfast table: books, games and clothes vouchers. All the stuff she'd expected but she pretended to be surprised. All the time she was thinking of soft-spoken Mister Mayhem who would be arriving later in the evening. An aunt dropped by and commented on her made-up eyes.

'Oh, you are grown up, Busy. Such beautiful eyes.'

Then an uncle.

'Oh, Busy. Someone will be losing their heart. So beautiful.'

She was embarrassed. She'd never thought of herself

in that way. She went to look in the mirror again. Maybe they were right. She smiled. She wasn't a kid any more. The birthday was…a transformation. A change into something different.

She saw it on the faces of her friends as they arrived – how their eyes widened, how they 'wowed!' and hugged her like she was a lost friend. Or a new friend. Everyone wanted to take a selfie with her. Was it her or was it just them? Did they seem younger? More child-like?

The noise levels grew as her 'gang' checked out the presents, holding up t-shirts, skirts. Just glimpsing at the books before they put them down quickly. The music was turned up louder so they all had to talk louder. Busy's parents brought in the food: snacks, pizzas and fizz. Hands reached in, mouths opened, the talking muffled. Busy didn't eat. She was wondering when Mister Mayhem would arrive. It might be a disaster. He would be terrible and her friends would laugh at her. Maybe he wasn't coming. Something had happened and he couldn't get here. Maybe that was a good thing. She felt a flutter of relief.

Ding! Dong!

Her heart lurched. Mister Mayhem. She grinned wildly at her friends.

'It's our special guest!' Busy announced, not sure if she could say, 'Mister Mayhem, the Clown,' for fear of the others' reactions.

'Mister Mayhem, the Clown!'

She had to go for it. She had to just do it. She wrenched the door open and there was…Mister Mayhem.

A howl of laughter went up. He WAS a clown! A full outfit of white and black silk with black bobble buttons down the front. The legs – the spangled pattern of the Harlequin. Big shoes which flapped as he walked. A battered hat with a flower in it. And a painted face – made weird by one half painted with a huge red smile and the other with a down-turned grimace and a painted tear in the corner of the eye. Mister Mayhem was a mixture of every clown –

Arlecchino, the tramp, the white-faced clown, the clown who cried…

Mister Mayhem bowed deeply. Busy bowed back automatically. Everyone laughed. Mister Mayhem tugged his hat off and held it towards Busy. As she went to take it, the flower squirted water all over her shoulders. She jumped back. Again everyone laughed. After a moment in which she wanted to tear the hat from his hands and crunch it into his smiling, crying face, she laughed too.

The clown stepped into the room, offering to shake hands with everyone. Sometimes he thumbed his nose, sometimes made a rude farting noise, sometimes something in his hand sent a shock into the other's hand and they yelped. He didn't stop: making jokes about their names, prodding them with a tickling stick, pulling eggs from behind their ears, making them look away then sticking his finger in their ears. Busy stood and admired the way Mister Mayhem, well, attacked

the room. Everyone was laughing and shrieking. Occasionally he would look at Busy and wink one painted eye. As if to say, 'You're safe from this…for now.'

Mister Mayhem went back to the door, waved goodbye and exited. The friends looked at each other and fell silent. Then they looked at Busy.

'Is that it?'

She shrugged. Surely not. Just a couple of minutes. Then came a sharp knock at the door. Busy ran to open it and there was Mister Mayhem with big, bright red box on little wheels which he pushed forwards, its wheels clunking and squeaking comically.

'My box of tricks!' he declared with a flourish of his arm.

And he settled it in the middle of the room and stood behind it as if he were serving ice creams. He beckoned everyone forward and signalled for them to sit. Busy stood at the back, unsure now of what was to come.

Mister Mayhem opened the lid of the box, pretending that it was a great weight, reached in and reached in, deeper and deeper. A desperate look passed over his face. Then a smile. He drew out a saw – stared at it and put it back, shaking his head. Next a pair of pliers. More puzzlement. Next a long-bladed knife with false red paint. The party guests squeaked as he brandished it over their heads. Then a pack of cards.

Busy was secretly pleased – this was all a little bit odd, a bit weird. She liked that. He did card tricks – guessing which cards had been selected, finding missing cards in people's pockets, making cards disappear into thin air. Everyone applauded. He was good.

After bowing again, he reached into the red box and drew out a handful of watches, some purses and a couple of mobile phones. He placed them on top of the box.

'They're all yours!'

Everyone looked puzzled then looked at their bare wrists, patted their empty pockets. They were indeed theirs. He was a clever pick-pocket. They all laughed again and were relieved when he handed them back, one by one. Slyly, Busy checked her watch and phone were still in her possession. They were. She laughed loudly. She was getting special treatment.

'How did he do that?' became the catch-phrase for the evening?

Then came juggling. Balls, tubes, phones, shoes – anything that came to hand. Two, three, four, five things, whirling, spinning, soaring, dropping – caught behind him, caught low just before they hit the ground, sideways, above his head – a dizzying pattern of spinning things. And in the middle of it, he winked at Busy again, as if to say, 'This is our show, Busy. You're part of this.' She smiled back, unsure but pleased, warmed by his…drawing her in.

Mister Mayhem closed the lid of the box. He looked from one to another. They stared back in anticipation. Busy watched the others. Yes, they were still children. But Mister

Mayhem knew she was different.

'And now a party game! Special hide and seek!'

The guests squeaked with delight. Busy smiled in her superior way.

'I'm going to put the lights out and you have to hide. If I find you, you come and sit down here in the dark, on the rug. Those are the rules. Okay?'

'Okay!' they chorused.

Mister Mayhem located the light switches. 'One! Two! Three!'

And suddenly the room was plunged into darkness. Shadowy shapes shuffled across the floor, bumping into each other, snuffling with laughter. Doors creaked open then shut. Curtains flapped as guests found their hiding places. Then there was quiet and stillness.

'I'm coming to get you.'

A few suppressed giggles. Mister Mayhem padded around the room, the corridors, declaring 'Found!' as he touched each guest. Gradually the rug in the centre of the room filled.

'Move into a circle around the rug.'

Shuffling. Bumping. More giggles.

'Now I have a special announcement to make. The birthday girl isn't here anymore.'

A few quizzical sounds. What on earth…?

'No. But I'm going to pass her around to you. In bits.'

Laughter.

'Here are her eyes.'

Hands took the objects and there were loud squeals of disgust.

'Yeeuuuch!'

'It's a lychee!'

'Ooooo-eeeeerrr. Busy's eyes! Ha ha!'

The objects moved round the circle.

Mister Mayhem spoke again, his voice a little harder.

'And this is Busy's tongue.'

He bent down and pushed another object into a girl's hand. She screeched in disgust and thrust it into another hand next to her.

'It's just a wet sponge, you idiot,' cried another.

More laughs. More screams.

Mister Mayhem's voice was louder this time making the guests duck their heads.

'And this Busy's heart!'

The squeals were of delight, shock, horror. All mixed with laughter.

'It feels real!'

'Oh my God!'

'Oooh…Busy's given her heart away.'

'Oh, shut up, clever pants.'

The squeals got louder.

'Yuuuch. It's really sticky.'

At that moment, the door rattled and opened. In the doorway stood Busy's mother and father, holding the birthday cake with the thirteen blazing candles.

'Don't come in!' shouted Mister Mayhem with panic in his voice.

But the flickering, golden light from the candles was revealing the circle of guests, now staring down at the scarlet pools and stains on their hands and clothes, the unmentionable red lumps scattered around them. That is when the wailing began and Mister Mayhem's voice singing.

'Happy birthday to you…Happy birthday to you…Happy Birthday dear Busy…'